The Interstellar Medium

NATO ADVANCED STUDY INSTITUTES SERIES

Proceedings of the Advanced Study Institute Programme, which aims at the dissemination of advanced knowledge and the formation of contacts among scientists from different countries

The series is published by an international board of publishers in conjunction with NATO Scientific Affairs Division

A	Life Sciences	Plenum Publishing Corporation
B	Physics	London and New York
C	Mathematical and Physical Sciences	D. Reidel Publishing Company Dordrecht and Boston
D	Behavioral and Social Sciences	Sijthoff International Publishing Company Leiden
E	Applied Sciences	Noordhoff International Publishing Leiden

Series C – Mathematical and Physical Sciences

Volume 6 – The Interstellar Medium

The Interstellar Medium

Proceedings of the NATO Advanced Study Institute held at Schliersee, Germany, April 2–13, 1973

edited by

K. PINKAU

Max-Planck-Institut für Physik und Astrophysik
Institut für extraterrestrische Physik, Garching bei München, Germany

D. Reidel Publishing Company

Dordrecht-Holland / Boston-U.S.A.

Published in cooperation with NATO Scientific Affairs Division

Library of Congress Catalog Card Number 73–91208

ISBN 90 277 0417 1

Published by D. Reidel Publishing Company
P.O. Box 17, Dordrecht, Holland

Sold and distributed in the U.S.A., Canada, and Mexico
by D. Reidel Publishing Company, Inc.
306 Dartmouth Street, Boston, Mass. 02116, U.S.A.

Printed in the Netherlands by D. Reidel, Dordrecht

TABLE OF CONTENTS

PREFACE

In recent years it has become apparent that contributions to our knowledge about the interstellar medium are made by practically all forms of astronomy ranging from radio- to gamma ray observations, and from cosmic ray measurements. It was thus thought fruitful to arrange for an interdisciplinary meeting of astronomers from the various fields of investigation, and of high energy astrophysicists. This meeting took place at Schliersee (Bavaria) from April 2 to 13, 1973. Lectures and some of the discussions held during that Advanced Study Institute are reproduced here.

Unfortunately, no manuscripts of the two lectures about infrared and cosmic ray observations were received and these are thus not available here. However, it was thought to be more important to proceed now with the publication.

The organisers are very grateful to Miss H. Eichele for her technical help during the meeting. The organisation of the Institute and the publication of the proceedings would have been impossible without the very great efforts and help of Mrs. M. Brunner and Mrs. D.Preis.

I would also like to express my gratitude to my colleagues K.W. Michel and S. Drapatz for their many contributions towards the success of the meeting.

Garching, October 10, 1973 K. Pinkau

Organisation: K. Pinkau, K.W. Michel, S. Drapatz

Max-Planck-Institut für Physik
und Astrophysik, Institut für
extraterrestrische Physik, Garching

Advisors: L. Biermann

Max-Planck-Institut für Astrophysik, München

R. Lüst

Max-Planck-Institut für extraterrestrische
Physik, Garching

J. Lequeux

Observatoire de Meudon

P.G. Mezger

Max-Planck-Institut für Radioastronomie, Bonn

M. Rees

Astronomical Institute, Cambridge

J. Trümper

Astronomisches Institut, Tübingen

We are very grateful to the Scientific Affairs Division
of NATO and to the Max-Planck-Society for sponsorship
of this conference.

LIST OF PARTICIPANTS

Lecturers

Carruthers, G.R., Washington
Dilworth, C., Milano
Habing, H.J., Leiden
van de Hulst, H.C., Leiden
Kahn, F.D., Manchester
Lequeux, J., Meudon
Mezger, P.G., Bonn
Pottasch, S.R., Groningen
Reeves, H., Paris
Shu, F.H., Stony Brook
Trümper, J., Tübingen
Woolf, N.J., Minneapolis

Participants

Andriesse, C., Groningen
Barsuhn, J., Bonn
Beale, J.S., Genf
Bedijn, P.J., Leiden
Bennett, K., London
Bibring, J.P., Orsay
Biermann, L., München
Billing, H., Manchester
de Boer, K.S., Groningen
Bonetti, A., Firenze
Brand, P., Edinburgh
Bussoletti, E., Meudon
Carson, P., Belfast
Cavallo, G., Bologna
Chaisson, E.J., Cambridge, USA
Charles, P.A., Dorking
di Cocco, G., Bologna
da Costa, A., Manchester
Danks, A.C., Mons
Dickinson, G.J., Durham City
Dopita, M., Manchester
Drapatz, S., Garching
Duin, R.M., Leiden
van Duinen, R., Groningen
Duraud, J.P., Orsay
Elliott, K.H., Manchester
Emerson, J.P., London
Flower, D., Meudon
Frisch, P., Chicago
Froeschle, Ch., Nice

Gomez-Gonzales, J., Meudon
van Gorkom, J.H., Groningen
Grün, E., Heidelberg
Guélin, M., Meudon
Haramundanis, K., Cambridge, USA
Havnes, O., Blindern
Higdon, J., Los Angeles
Holmes, J.A., Oxford
Isobe, S., Heidelberg
Israel, F.P., Leiden
Jørgensen, H., Copenhagen
Kaftan-Kassim, M.A., Albany
Kisselbach, V.J., Lindau/Harz
Lada, Ch.J., Cambridge, USA
Launay, J.M., Meudon
Lee, P., Baton Rouge
Lee, T.J., Edinburgh
Lucas, R., Meudon
Macchetto, F., Frascati
Mayer-Haßelwander, H.A., Garching
Mebold, U., Bonn
Michel, K.W., Garching
Mills, R.M., Falmer
Moyano, C., Garching
Muzumdar, A.P., Bristol
Nagarajan, S., Bruxelles
Natta, A., Frascati
Norman, C.A., Oxford
d'Odorico, S., Asiago
O'Neill, A., Falmer
Panagia, N., Ithaca
Payne-Gaposchkin, C., Cambridge, USA
Peterson, F.W., Charlottesville
Pinkau, K., Garching
Poulakos, C., Athens
Reina, C., Milano
Robinson, B.J., Sydney
Röser, H.J., Göttingen
Roueff, E., Meudon
Schlüter, A., Garching
Schmidt, W., Garching
Schönfelder, V., Garching
Schwehm, G., Bochum
Shukla, P., Garching
Sieber, D., Bonn
Smeding, A.G., Groningen
Tenorio-Tagle, G., Manchester
Tscharnuter, W., Göttingen
Voges, W., Garching

Weliachew, L., Meudon
Whitworth, A.P., Leiden
Winkler, K.H., Göttingen
Winnewisser, G., Bonn
Woodward, P.R., Berkeley
Yorke, H., Göttingen

WHERE IS THE INTERSTELLAR MEDIUM?

H.C. van de Hulst

Sterrewacht te Leiden
Leiden, The Netherlands

Abstract: As a historical introduction, this paper reviews the various distinctions successively introduced to locate the interstellar matter more precisely.

A. INTRODUCTION

The history of science is like a wiggly curve. It has inadvertent bumps and interruptions, as if someone with a not too expert hand had tried to draw a smooth curve. The real trend is seen only over a sufficiently long time interval. Too often young authors start the history of their subject 3 years ago. This may be justified in a field where recent improvements in instrumentation or in understanding are so great that all earlier work can be ignored. Interstellar matter typically is not in this exceptional case. Many present "theories" are based on much earlier tentative interpretations of observations that by themselves were not too definite. For that reason, a systematic review must at the same time be historical. The contents of the words gradually change with their repeated use.

B. THE LEADING QUESTION

Where is the interstellar medium? The answer seems obvious from the literal meaning of the word: between the stars. But from the onset this word has rather been used for its negative meaning: <u>not in</u> the stars, in contrast to the media studied in the classical divisions of astrophysics, viz. stellar atmospheres and stellar interiors.

We shall now explain, roughly in historical order, most of the distinctions that have been introduced to state more precisely where interstellar matter is.

1. Intergalactic - interstellar - interplanetary

The scale difference between the regions indicated by these words is so enormous that there would not seem to be any problem of confusion ever. Yet the border lines have not always been clearly seen. Up to 1940, one finds papers linking the chemical composition of interstellar grains to that of meteorites. This link seemed inescapable because - before the radar measurements - the hyperbolic velocity of certain meteors seemed "well established". About 1949, the major issue in cosmic ray physics was whether cosmic rays were galactic (i.e. interstellar) or pervaded the entire universe. By now, intergalactic matter has become a subject of its own right. And so has interplanetary matter with solar wind, cosmic ray modulation and other topics of great interest.

2. Interstellar - circumstellar

The stationary absorption lines discovered in 1904 clearly showed the presence of gas that did not participate in the orbital motion of the double star. This might point to a general interstellar medium but a circumstellar cloud roughly 10 times larger than the orbit might do as well. For the stronger lines abundant statistical data now exist to show that they are truly interstellar. But throughout the 30's and 40's there remained a suspicion that the molecular lines, which had been observed in only a few stars, might after all be circumstellar. The word "circumstellar" remained a bleak word, until the recent discoveries, by very-long-baseline radio observations and by infrared observations, that objects of the size of the solar system do indeed exist. We have now entered the stage in which further distinctions inside the class of circumstellar objects must be made. These will refer first of all to the evolutionary status of the object: whether it is a still contracting protostar or a fully evolved object.

3. Nebulae - interstellar matter

In the old times (say 1935), when the 18th century indiscriminate word "nebulae" had already been subdivided into numerous types, it remained customary to talk about nebulae <u>and</u> interstellar matter. The similarities in the physics of these two things were clearly seen but they were thought of as distinct astronomical objects. The typical densities (number of H-atoms, neutral or ionized) were:

gaseous nebulae	planetary nebulae	$= 10^4$ cm^{-3}
	diffuse emission nebulae	$= 10^2 - 10^3$ cm^{-3}
	interstellar medium	$= 10^0 - 10^1$ cm^{-3}

The present feeling is that the large diffuse emission nebulae form a passing phase in the evolution. They will not stay together and will gradually dissolve. The very bright ones, like the 5 or 6 which are in Messier's catalogue, rightly carry proper names as individuals. However, in mapping the sky down to smaller and smaller values of the emission measure it becomes progressively more difficult to name individual objects. It soon becomes a matter of taste and, finally, this task is completely futile.

4. HI and HII-regions

This is one of the most tricky distinctions, historically. It entered the literature in a well-defined year (1939). The theoretical concept was simple. A hot star placed somewhere in the hydrogen gas can keep only a limited region ionized. This region (HII-region, Strömgren sphere) must be spherical, for symmetry, if the gas is homogeneous. Its boundary must be rather sharp because the small free path of an ionizing quantum in the neutral gas causes a fairly sudden transition from 100% to 0% ionization . Strömgren's study was inspired by the recent discovery, by Struve and coworkers, of roughly circular regions of faint Hα emission.

The tricky point enters here. At the time it seemed natural to identify the theoretical concept and the observed fact and this caused the same word, HII-region, to be used for both. At present, the identification is still roughly correct, but not quite. Many of the faint emission regions may be mass-limited instead of ionization-limited in size. Yet the name HII-regions sticks with them. It is for such objects strictly a misnomer but established terminologies cannot easily be changed. So the next best thing is to warn students against possible confusion.

Later variations on the theory involved the calculation of the depth of an ionized rim to a gas cloud and of the size of the ionized region in a spotty medium consisting of separate dense clouds. Also, if the ionizing source is an X-ray source, the freepath is much longer and the boundary less sharp.

5. Medium - cloud

Both words are plain-language words that may carry many different meanings. "Medium" has about it the connotation of being smoothly distributed, or about homogeneous. "Cloud" has the connotation of a striking difference of density, or other property, and a rather abrupt transition between cloud and surrounding. These descriptive words have entered into the literature on interstellar matter along two entirely separate routes.

(a) The nearest dust clouds are visible to the naked eye as "dark clouds". These and other, less striking, ones have been studied extensively by star counts and Wolf diagrams from about 1910. A general dust medium was suspected by many early authors but Trumpler's study of 1932 usually is quoted as the first proof of its existence. Initially these subjects remained separate. Only about 1940, through the more reliable statistics available in the photoelectric B-star colors of Stebbins' group, the conviction broke through that the "general extinction" was the net effect of all "clouds". One of the consequences was that the observed average extinction coefficient in the galactic plane, 0.7 or 1.0 mag/kpc, was recognized as a biased average of fairly clear regions, and that a "fair" average should be 2.0 mag/kpc or even higher.

(b) Gas clouds appeared in the literature in the interpretation of Adams' Mt Wilson coudé spectra of interstellar ionized calcium lines. The resolution then available, about 7 km/sec, did not suffice to separate all line components but separate components with deviating velocities were quite clearly seen. The reason why Adams called them "clouds" is that a clear separation in velocity-space is about unthinkable without a similarly clear separation along the line of sight. This did - and does - tell nothing about the shape. Therefore, "shells", "curtains", "streams" would be equally admissable guess-words for this phenomenon. The present situation both for radio and optical spectroscopy will be discussed in later talks.

Comparison (a)-(b): In comparing the early history of the study of dust clouds and gas clouds we see that they have nothing in common and that the basic observational facts are of a quite different nature. Therefore, if at all the pursuits (a) and (b) give similar results for cloud size or cloud density, this should be a reason for surprise. For the past 25 years the most popular cloud model had a density of 10 H-atoms per cm^3 and a diameter of 10 parsec. These numbers were first introduced as "a convenient working model", or as "an attempt of an estimate" but have by long use solidified into specifications of the "standard cloud". Numbers of this order can be derived in various uncertain ways from gas clouds alone (via ionization equilibrium and abundance ratios or via emission measures) or from dust clouds alone (via angular size and distance) or, more easily via a combination of both (number of clouds cut per line of sight $\sim$ 10 kpc^{-1}, fraction of space occupied by clouds $\sim$ 0.02). However, in detailed research problems the student should remain on his guard against a too easy identification of either gas clouds or dust clouds with this "standard".

6. Disk-halo

In the historical review we now have passed the time that most students in this course were born. At the IAU-meeting, Dublin 1955, the galactic halo was the hot subject. Until that time, any discussion on interstellar matter had referred to the disk, because intrinsically bright stars are necessary for diffuse emission nebulae, for extinction measurements, and for the observations of absorption lines. Roughly, the disk was known to extend from the galactic plane about 50 or 100 parsec either way.

The halo concept arose first of all out of radio astronomy. It seemed unmistakable that continuous non-thermal radio emission was seen in a spherical or spheroidal distribution extending many kiloparsecs above the galactic plane. Then there must also be gas, supporting the electric currents, supporting the magnetic fields, supporting the radio emission. Magnetic fields were also necessary to keep the more energetic cosmic rays inside the galaxy. And, finally, Spitzer's speculation of a hot tenuous intercloud gas providing the pressure preventing the interstellar clouds to expand led to the theoretical prediction of a halo.

Progress in this field has been remarkably little. Other galaxies do not show the prominent haloes expected at that time. Again, I must leave it to other speakers to take up the subject for an up-to-date review.

7. Arm - interarm

In 1952, it became known as a firm fact, both from optical data (OB-associations and HII-regions) and from radio data (21-cm line) that our galaxy possessed arms. The exact form was debatable but, generally, they were welcomed as spiral arms. Who expects that from that time on all earlier distinctions fell into place, as referring either to an arm or an interarm region, will be disappointed. Such attempts were made only very recently. For it is now generally thought that the arms have not only a geometrical but also an evolutionary significance. Matter must during the course of time stream through the interarm region, be retarded while crossing the arm, and so on.

One reason why it was hard to link the new arm - interarm distinction to the body of earlier knowledge about interstellar matter probably is the different scale. Traditional interstellar studies were "local", quite often referring to $r < 500$ pc or 1 kpc, rarely going to 2 kpc. On the other hand, the conversion from velocity to distance, which is the principle used for mapping spiral arms from 21-cm observations, is based on differential rotation. This comes out clearly above the "noise" of the random cloud motions only at distances well beyond 1 kpc.

8. Globules - cloudlets - compact regions - rims - wisps

Although several statistical approaches lead to a "classical cloud" of about 10 parsecs diameter as representative, any direct observation permitting a better resolution shows indeed smaller detail. A variety of descriptive words have been introduced for these details.

There is no accepted hierarchy among them. Yet it is clear that all of these details must have an origin in the gas dynamics of the interstellar medium, often complicated by the ionizing radiation of stars and sometimes by the presence of magnetic fields. Details that are sufficiently small and dense and devoid of internal motions may start contracting into protostars. A great deal of the interest given to these small details derives just from the fact that at those places the stars may be formed that provide the new energy to keep the interstellar medium churned up.

C. PHYSICAL PROPERTIES

In the oral presentation, the preceding lecture (section B) was followed by one reviewing systematically how the physical properties (density, temperature, degree of ionization, magnetic field) and the chemical composition of the various objects can be determined. Since this subject is treated in many textbooks and will anyhow be covered in some of the lectures that follow, this presentation is not reproduced.

The great variety of places where interstellar matter can be observed in various forms and by various methods is indeed bewildering. The distinctions made one by one and discussed in the preceding section, have made it quite impossible still to speak of *the* density or *the* temperature of interstellar matter. The well-meant but somewhat schematic stories of the 1930's about what interstellar matter *might* be like have been largely replaced by specialized studies about what it *is* like. The facts are, as usual, annoyingly complex but at the same time much more fun than the earlier fiction. Just go and spend some time with some sheets of the beautiful Palomar Schmidt sky atlas and a good magnifier.

D. REFERENCES

It would take hundreds of references and scores of illustrations to support the preceding text properly. Perhaps someone will find the inspiration to do this some time, although he must realize that, with the many symposia and visits, the most important events in the history of scientific ideas may not be found black-on-white.

Among the older original papers I would recommend

Eddington's Bakerian lecture (Eddington, 1926), Oort's George Darwin lecture (Oort, 1946), and Strömgren's discussion on the ionization (Strömgren, 1948). A review of the gas dynamics of interstellar matter, with a quick guide through half a dozen earlier symposia and with emphasis on the large-scale energy balance and mass balance is found in van de Hulst (1970). The best book on interstellar matter undoubtedly is Spitzer's (1968), but it may be interesting to use Kaplan and Pikelner (1970) at its side. For all further matters, some of which may be a few years old and others not yet published at all, I refer to the further lectures in this book.

REFERENCES

Eddington, A.S., Proc.Roy.Soc.A 111, 423, 1926.

van de Hulst, H.C., in "Interstellar gas dynamics",IAU-Symposium No. 39, ed. H.J. Habing, Reidel, 1970.

Kaplan, S.A. and Pikelner, S.B., "The interstellar medium" Harvard Univ. Press, 1970.

Oort, J.H., Mon.Not.R.A.S. 106, 159, 1946.

Spitzer, L., "Diffuse matter in space", Interscience,1968.

Strömgren, B., Astrophys.J. 108, 242, 1948.

RADIO OBSERVATIONS

P.G. Mezger

Max-Planck-Institut für Radioastronomie
Bonn, W-Germany

I. INTRODUCTION

Figure 1 shows a schematic cross section through our galaxy, perpendicular to the galactic plane. Stars are concentrated towards the galactic center, where they form the "nuclear bulge". Their numbers gradually taper off with increasing distance from the galactic center. The interstellar gas forms a flat layer, whose thickness in latitude increases with increasing distance from the galactic center. Globular clusters are the oldest stellar systems and therefore must have been formed in the early evolutionary stages of the galaxy.

The upper half of the diagram shows the surface densities (i.e. projected onto the galactic plane) of the various constituents. Whereas the surface density of stars increases rapidly with decreasing distance from the galactic center, the surface density of atomic hydrogen attains a very flat maximum between 15 and 4 Kpc. Stars and interstellar matter rotate around the center of gravity of the galaxy. At the distance of the sun this velocity is 250 km/sec. It takes the sun 2.5×10^8yr to complete one rotation around the galactic center. The rotation curve of the galaxy, as obtained partly from 21-cm observations, partly from model calculations, is shown in the upper right side of the diagram. The total mass of the galaxy derived from this rotation curve is

$$M_{tot} \sim 1.8 \times 10^{11}\ M_\odot$$

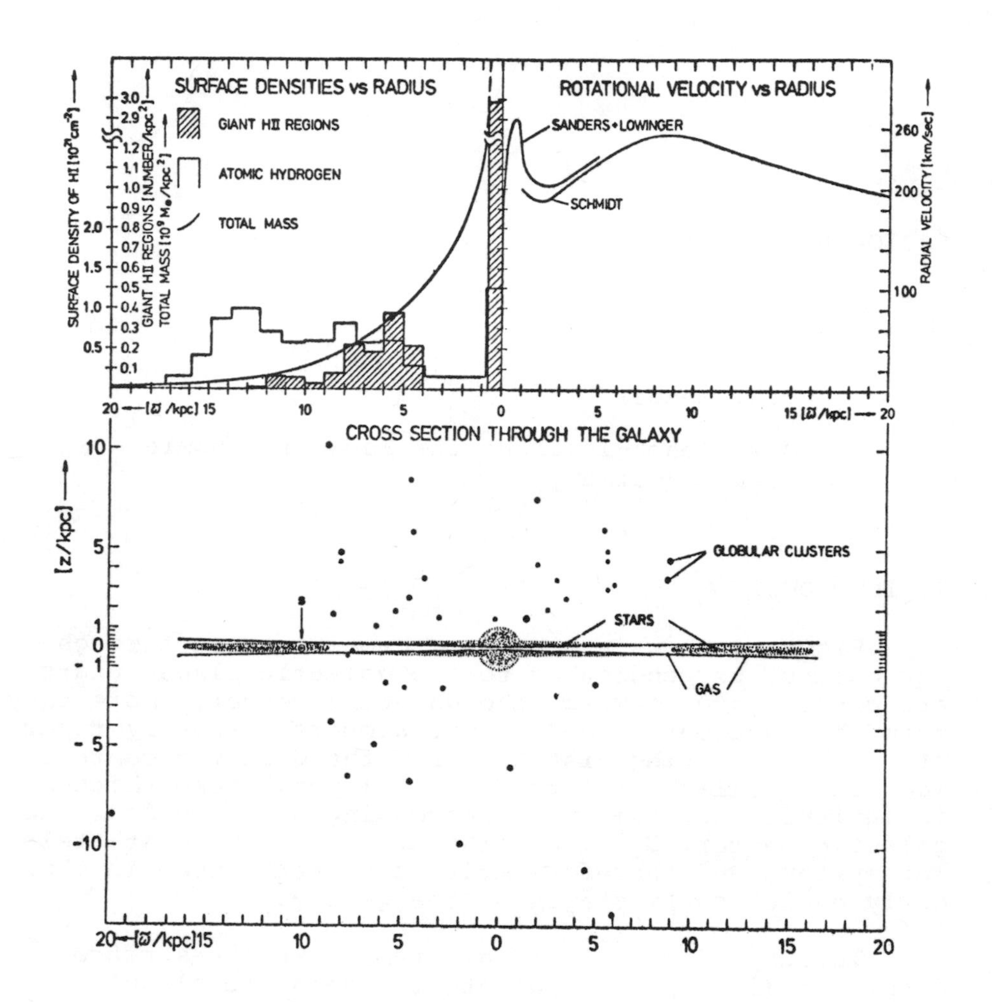

Fig. 1: Distribution of stars and interstellar matter in the galaxy. The lower part of the diagram is a cross section perpendicular to the galactic plane. The upper left part of the diagram shows the column density of stars, hydrogen gas and HII regions as a function of the distance $\tilde{\omega}$ from the galactic center. The rotational (orbital) velocity is given as a function of $\tilde{\omega}$ in the right part of the upper diagram.

The total mass of atomic hydrogen, obtained from integration of 21-cm line profiles, is $\sim 3 \times 10^{-2} M_{tot}$. Considering the presence of ^4He and of "unseen matter" (unseen, because $\tau_{21cm}>1$ or H is tied up in H_2-molecules), I estimate that the mass ratio of the interstellar matter to the total, $M_{IM}/M_{tot}\sim 0.1$ for our galaxy. Note, however, that this ratio varies strongly with distance from the galactic center.

The main constituents of the interstellar gas (and their relative abundance, by number) are H(1), ^{4}He(10^{-1}), C, N, O, Ne ($\sim 10^{-4}$) and Mg, Si, S, Fe($\sim 10^{-5}$). H and ^{4}He account for 70% and 28%, respectively; all elements heavier than ^{4}He account for 2 % of the total mass of interstellar gas. Dust particles, with typical radii from $a\sim 0.15$ to 0.02μ, account for a mass fraction of ~ 1% of the interstellar matter. Extinction by these dust particles limits optical and UV observations in the galactic plane to the solar vicinity. The propagation of radio waves, on the other hand, is not impeded by dust particles. The large scale structure and integral characteristics of the interstellar matter are therefore primarily derived from radio observations.

Density and kinetic temperature of the interstellar gas vary by many orders of magnitude. The two-component structure, cool and dense cloudlets embedded in a hot and tenuous intercloud gas, appears to be a general characteristic of the interstellar gas. The very dense, cool and massive "dark" and "black" clouds, on the other hand, appear to be related to the large-scale structure of the interstellar gas, such as material and density-wave spiral arms. These dense and cool clouds appear to be the first evolutionary stages in the formation of stars out of the interstellar matter. Once O-stars are formed they will ionize a large fraction of the surrounding interstellar gas, forming the HII regions. Density and temperature ranges of the various components of interstellar matter are compiled in Table 1.

The partial ionization of the gas in the cloudlets and in the intercloud regions is maintained by cosmic rays and X-rays whereas, in dense clouds, photoionization of carbon provides most of the free electrons. Cloudlets and intercloud gas are investigated by means of the Hλ21cm line. In dense clouds the 21cm line becomes opaque; in addition, most of the atomic hydrogen becomes tied up in molecules, which form out of the most abundant elements H, C, N, O, Si and S. Many of these molecules have rotational transitions in the radio range, which, at present, provide the only means to investigate the physical state of dense clouds.

Component of Interstellar Matter	Densities/cm^{-3}	$T_k/^oK$	n_e/n_H	Radiation mechanism *)
Dark and black clouds Molecular clouds	$10^3 \le n_{H_2} \le 10^7$	5 - 60	$\le 2.3 \times 10^{-4}$	Molecular lines (E,A) C-recombination lines (E)?
Cloudlets	$10 \le n_H \le 10^2$	50 - 100	$\sim 10^{-3}$	H λ21cm (E,A), OH λ18cm (A) H_2CO λ6cm (A)
Intercloud gas	$10^{-2} \le n_H \le 10^{-1}$	10^3-5×10^3	$\sim 10^{-1}$	H λ21cm (E), free-free (A)
HII Regions	$n_e < 10^5$	5×10^3-10^4	>> 1	free-free (E); H- and He-recombination lines
Supernovae and Diffuse Relativistic Electrons				Synchrotron or Magneto-Bremsstrahlung

*) A = Absorption; E = Emission

Table 1: Physical State and Dominant Radio Radiation Mechanisms of Various Components of the Interstellar Gas.

A plasma emits in the radio range a free-free continuum (or Bremsstrahlung) and a recombination spectrum arising from transitions with principal quantum numbers $n \gtrsim 50$. At short wavelengths, this thermal continuum radiation is the predominant emission from galactic sources. At long wavelengths, the presence of a thermal plasma manifests itself primarily through its absorption of non-thermal emission. Attempts to detect continuum or recombination line emission from the partially ionized interstellar gas have provided, up to now, only ambiguous results; only some observations of free-free absorption have been unambiguously attributed to the partially ionized interstellar gas.

Apart from cosmic rays, interstellar space is permeated by cosmic (relativistic) electrons. These electrons, decelerated in the galactic magnetic field, emit synchrotron or Magneto-Bremsstrahlung radiation, which is responsible for the diffuse non-thermal galactic radiation. Synchrotron radiation is also responsible for the radio radiation by supernova remnants. In Table 1, the principal radiation mechanisms used for the investigation of the various components of interstellar matter are indicated.

The interstellar photon radiation consists of three components: (I) X-ray radiation, which appears to be primarily of extragalactic origin but may have a soft galactic component; (II) UV- and optical radiation from stars; (III) blackbody radiation corresponding to a radiation temperature of $T_{bb} = 2.7^{o}K$, which is the diluted and redshifted primordial fireball. Components (I) and (II) are largely responsible for the ionization of the interstellar gas, whereas the $2.7^{o}K$ blackbody radiation sets a lower limit to the kinetic and LTE excitation temperature.

This sets the stage for radio observations. In the following, I will review the principal characteristics of the various radio radiation mechanisms which, in subsequent lectures, will be used to derive the physical state of the interstellar gas.

II. OBSERVING TECHNIQUES

The basic radioastronomical observing instrument is a single-dish parabolic reflector telescope. The characteristics of such a telescope are fully known if its power pattern $f(\Theta,\Phi)$ or if at least its integral, the antenna solid angle

$$\Omega = \int_{4\pi} f(\Theta,\phi)\, d\Omega \tag{II.1}$$

is known. Unfortunately, $f(\Theta,\phi)$ can neither be measured nor computed for large radiotelescopes. One therefore measures the half power beam width (HPBW = width between half power intensity points), approximates the main lobe by a Gauss-function and obtains, for the main beam solid angle,

$$\Omega_m = \int_{\Omega_m} f(\Theta,\phi)\, d\Omega = 1.133\,\Theta_A^2 \tag{II.2}$$

A good approximation for the telescope HPBW is

$$\left[\frac{\Theta_A}{\text{arc min}}\right] = 4.176 \cdot 10^3\ \lambda / D \tag{II.3}$$

The beam efficiency

$$\eta_B = \Omega_m / \Omega \tag{II.4}$$

is usually obtained through the use of a standard calibration source. Typical values are $\eta_B = 0.7 - 0.9$.

Consider now a radiotelescope pointing at a radio source of surface brightness $B_\nu(\Theta,\phi)$. As always in astrophysics, we compare the surface brightness to that of a blackbody, i.e.

$$\left[\frac{B_\nu}{\text{W m}^{-2}\ \text{Hz}^{-1}\ \text{ster}^{-1}}\right] = \frac{2h\nu^3}{c^2}\ \frac{1}{\exp\left\{\frac{h\nu}{kT}\right\}-1} \simeq \frac{2kT\nu^2}{c^2} = \frac{2kT}{\lambda^2}\ \text{for}\ \frac{h\nu}{kT} << 1 \tag{II.5}$$

The "Rayleigh-Jeans" (RJ) approximation is used throughout the radio range, although

$$T >> T^* = h\nu / k = 4.8 \cdot 10^{-2} \left[\frac{\nu}{\text{GHz}}\right] \tag{II.6}$$

is not always fulfilled. In this case an RJ brightness temperature is defined by

$$T_b(RJ) = T^* \frac{1}{e^{T^*/T_b} - 1} \tag{II.7}$$

E.g. for $T_b = 2.7^\circ K$, $\nu = 115$ GHz, $T_b(RJ) = 0.9^\circ K$. Note that very few celestial objects - the moon and planets are examples - radiate like blackbodies and, thereby, have brightness temperatures which are independent of frequency. The radiotelescope receives the power

$$\left[\frac{P_\nu}{\text{W Hz}^{-1}}\right] = \frac{1}{2}\ \frac{\lambda^2}{\Omega} \int_{4\pi} B\nu\, f\, d\Omega = \frac{k}{\Omega} \int T_b\, f\, d\Omega$$

If the radiometer is now connected to a matched load of variable temperature, T_A, it receives the power $P_\nu = kT_A$. Thus, the power received by the antenna can be described by the equivalent antenna temperature

$$T_A = \frac{1}{\Omega}\int_{4\pi} T_b(\Omega)\, f(\Omega)\, d\Omega = \frac{1}{\Omega}\left\{\int_{4\pi-\Omega_m} \ldots d\Omega + \int_{\Omega_m} \ldots d\Omega\right\}$$

Most observations are made in such a way that the first term is constant or varies in a linear way during an observation and thus can be eliminated. Using (II.4), we obtain the following relation between the brightness temperature distribution in the main beam and the antenna temperature

$$\frac{T_A}{\eta_B} = \bar{T}_b = \frac{1}{\Omega_m}\int_{\Omega_m} T_b(\Omega)\, f(\Omega)\, d\Omega \qquad \text{(II.8)}$$

$\bar{T}_b$ is the main beam brightness temperature; most observations are given in units of this main beam brightness temperature. Let us consider a Gaussian distribution of the brightness temperature of HPW θ_s with T_b the peak brightness temperature. Then $\Omega_s = 1.133\,\theta_s^2$ and evaluation of eq. (II.8) yields

$$\bar{T}_b = \frac{T_A}{\eta_B} = T_b \frac{\theta_s^2}{\theta_A^2 + \theta_s^2} = \begin{cases} T_b & \text{for } \theta_s^2 >> \theta_A \\ T_b\left(\frac{\theta_s}{\theta_A}\right)^2 & \text{for } \theta_s^2 << \theta_A \end{cases} \qquad \text{(II.9)}$$

Especially for continuum sources, the flux density

$$\left[\frac{S_\nu}{\text{W m}^{-2}\,\text{Hz}^{-1}}\right] = \int_{\Omega_s} B_\nu \, d\Omega$$

is a useful quantitative measure which is not dependent on the characteristics of the telescope. Integration has to be extended over the solid angle Ω_s covered by the source. In radioastronomy one uses the unit

$$1 \text{ f.u.} = 10^{-26} \text{ W m}^{-2} \text{ Hz}^{-1}$$

or

$$1 \text{ m.f.u.} = 10^{-29} \text{ W m}^{-2} \text{ Hz}^{-1}$$

for convenience. One can show that flux density and main beam brightness temperature are connected by

$$S_\nu = \begin{cases} \frac{2k}{A} T_A & \text{for } \theta_s << \theta_A \quad \text{"point source"} \\ \frac{2k}{\lambda^2}\int_{\Omega_s} \frac{T_A}{\eta_B}\, d\Omega & \text{for } \theta_s \geq \theta_A \quad \text{"extended source"} \end{cases} \qquad \text{(II.10)}$$

Here

$$A = \lambda^2 \Omega = \lambda^2 \frac{\Omega_m}{\eta_B} \tag{II.11}$$

with Ω and Ω_m in steradians, λ in meters, is the effective antenna area. The power absorbed by an antenna from a plane wave of flux density S_ν is $P_\nu = (\frac{1}{2}) S_\nu$ with "1/2" accounting for the fact that the antenna receives only in one plane of polarization. The ratio of effective to geometrical antenna area

$$\eta_A = A/\pi (D/2)^2 \tag{II.12}$$

is the antenna efficiency; a typical value of $\eta_A \simeq 0.5$. Angular resolution of single dish telescopes is $\gtrsim 1'$; of aperture synthesis telescopes is $\gtrsim 1''$.

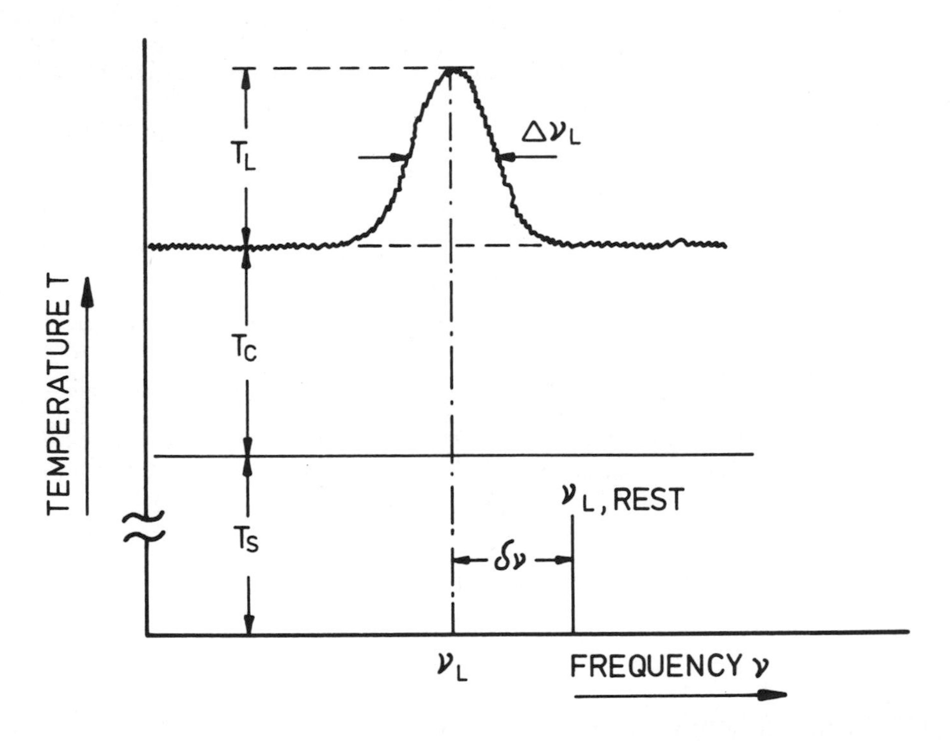

Fig.2: Schematic representation of an observed line profile. T_C represents the (frequency independent) continuum spectrum of the radio source. The scale of the system noise temperature T_S is compressed.

In the most general case, a line of antenna temperature T_L (Fig.2) is observed superimposed on a continuum source of temperature T_C, which is the sum of all radio continuum emission along the line of sight of the radio-telescope, including the universal 3°K blackbody radiation. Usually the center frequency ν_L of the observed line is shifted with respect to the laboratory frequency by an amount $\delta\nu$ as a result of Dopplershift, which is expressed as radial velocity of the emitting source relative to the sun (heliocentric velocity) or to the local standard of rest (LSR), the latter being defined by the standard solar motion of +20 km/s toward $l^{II} = 56.2°$ and $b^{II} = 22.8°$. Note that, in astronomy, a positive radial velocity means a redshift or recession of the emitting source, i.e.

$$V_r = -\frac{\delta\nu}{\nu_{L,rest}} \times c$$

with $c = 2.997\ 929 \times 10^5$ km/sec^{-1}. The total energy contained in the line profiles is measured by the integral $\int T_L\, d\nu$. In the case of a Gaussian line profile of HPW $\Delta\nu_L$ this quantity is $1.065\ T_L\, \Delta\nu_L$. The line profile is analyzed in a (multichannel or autocorrelation) spectrometer. The bandpass of an individual channel determines the frequency resolution of an observation; the total bandwidth covered by the spectrometer in a given observation is referred to as its (instantaneous) analyzing bandwidth.

III. RADIATION MECHANISMS AND EQUATION OF TRANSFER

(a) Line emission and absorption by a neutral gas

Consider a cloud of interstellar gas located between the sun and a source of continuum radio emission (e.g. an HII region or a SN remnant) characterized by its surface brightness T_C. In this cloud, we consider a specific atom (or molecule) with a transition from an upper (u) to a lower (ℓ) level. The frequency of this transition is

$$\nu_{u\ell} = \frac{E_u - E_\ell}{h}$$

Emission or absorption of the corresponding spectral line can be described in terms of the optical depth $\tau_L(\nu)$ and the excitation temperature T_{ex}. The brightness temperature at a frequency inside the line profile is

$$T_L' = T_{ex}(1 - e^{-\tau_L}) + (T_c + T_{bb})\, e^{-\tau_L}$$

Here, we have added the $T_{bb} = 2.7°K$, blackbody temperature to the continuum brightness temperature T_c of the (optically thin) radio source. Outside the line, we observe the unattenuated continuum temperature $T_C' = T_{bb} + T_C$. The observed line brightness temperature is, therefore,

$$T_L = T_L' - T_c = \{T_{ex} - (T_c + T_{bb})\}(1 - e^{-\tau_L}) \tag{III.1}$$

If $T_{ex} > (T_C + T_{bb})$ the line is seen in emission; if $T_{ex} < (T_C + T_{bb})$ the line is seen in absorption. The excitation temperature is defined by

$$\frac{n_u}{n_\ell} = \frac{g_u}{g_\ell} e^{-h\nu/kT_{ex}} \tag{III.2}$$

where n_u and n_ℓ are the number of atoms (or molecules) with an electron in the energy levels E_u and E_ℓ, respectively, and g_u and g_ℓ are the statistical weights. In Thermodynamic Equilibrium (TE) T_{ex} is the same for all levels. In interstellar space, however, deviations from TE are the rule. However, statistical equilibrium must still apply; i.e. the number of (spontaneous and induced) downward transitions must equal the number of upward transitions by absorption.

$$n_u \{A_{u\ell} + u_\nu B_{u\ell}\} = n_\ell u_\nu B_{\ell u} \tag{III.3}$$

Here, $u_\nu = \frac{4\pi B_\nu}{c}$ is the radiation density with B_ν given by eq. (II.5). Between the Einstein-A- and B-coefficients the following relations hold:

$$g_\ell B_{\ell u} = g_u B_{u\ell} \tag{III.5}$$

$$g_\ell B_{\ell u} = \frac{c^3}{8\pi h\nu^3} g_u A_{u\ell} \tag{III.6}$$

and the net absorption coefficient per cm

$$\int_0^\infty \alpha' d\nu = \frac{h\nu}{c} \{B_{\ell u} n_\ell - B_{u\ell} n_u\} = \frac{h\nu}{c} B_{\ell u} n_\ell \underbrace{\left\{1 - \frac{g_\ell n_u}{g_u n_\ell}\right\}}_{\sim \frac{h\nu}{kT_{ex}} \text{ for } kT_{ex} >> h\nu} \tag{III.7}$$

Combining (III.6) and (III.7)

$$\int_0^\infty \alpha' d\nu = \frac{c^2 h}{8kT_{ex}\nu} \frac{g_u}{g_\ell} A_{\ell u} n_\ell \tag{III.8}$$

and the optical depth

$$\int_0^L ds \int_0^\infty d\nu\, \alpha' = \int_0^\infty \tau_\nu\, d\nu = \frac{c^2 h}{8\pi k\nu} \frac{g_u}{g_\ell} A_{\ell u} \frac{\int_0^L n_\ell\, ds}{T_{ex}} \tag{III.9}$$

Here

$$N_\ell = \int_0^L n_\ell \, ds \qquad \text{(III.10)}$$

is the column density of atoms or molecules with an electron in the energy level ℓ. The optical depth per Hz bandwidth is

$$\tau_\nu = \phi_\nu \int_0^\infty \tau_\nu \, d\nu \qquad \text{(III.11)}$$

with ϕ_ν the profile function, normalized such that $\int_0^\infty \phi_\nu \, d\nu = 1$. For a Gaussian profile of HPW $\Delta\nu_L$ between half intensity points

$$\phi_\nu = \frac{0.9394}{\Delta\nu_L} \exp\left\{ \frac{(\nu_L - \nu)^2}{0.6006 \quad \Delta\nu_L^2} \right\} \qquad \text{(III.12)}$$

Gaussian profiles are the result of statistical Doppler broadening due to thermal motions or microturbulence

$$\Delta\nu_L = \frac{2\nu_L}{c} \left[\ln 2 \left\{ \frac{2kT_K}{m_A} + \frac{2}{3} \langle v_t^2 \rangle \right\} \right]^{1/2} \qquad \text{(III.13)}$$

Here, m_A is the mass of the emitting atom (or molecule), $\langle v_t^2 \rangle_{rms}$ the rms velocity of the microturbulence.

(b) Synchrotron Radiation

A relativistic electron of energy $E = m\, c^2/(1-v^2/c^2)^{1/2} \gg m\, c^2$, decelerated in a magnetic field of strength $H_\perp$ perpendicular to its velocity vector, moves in a spiral. Its angular velocity of rotation $\omega_{H_\perp}$ and the radius r of its orbit are

$$\omega_{H_\perp} = \frac{eH_\perp}{mc} \frac{mc^2}{E} \quad \text{and} \quad r = \frac{c}{e} \frac{E}{H_\perp}$$

It radiates in a narrow cone of angle $\phi = m\, c^2/E$. An observer located in the plane of orbit sees a sequence of short-lived pulses of duration

$$\Delta t = (1 - \frac{v}{c}) \frac{r\phi}{c} \simeq \frac{mc}{eH_\perp} \left(\frac{mc^2}{E}\right)^2$$

which repeat at the frequency $\omega_{H_\perp}$. The decomposition of these periodic pulses gives a large number of harmonics of $\omega_{H_\perp}$ which, for $\Delta t \ll \omega_{H_\perp}^{-1}$ yield, for most practical purposes, a continuous spectrum $P(\nu,E)$ with peak emission at $\nu_m = 0.3\nu_c$, where

$$\omega_c = 2\pi\nu_c = \frac{3}{2\Delta t} = \frac{3}{2}\frac{eH_\perp}{mc}\left(\frac{E}{mc^2}\right)^2$$

or+)

$$\left[\frac{\nu_c}{\text{MHz}}\right] = 16.0\left[\frac{H_\perp}{\text{Oe}}\right]\left[\frac{E}{\text{MeV}}\right]^2 \qquad \text{(III.14)}$$

As a next step, one computes the emission coefficient

$$\varepsilon(\nu) = \frac{1}{4\pi}\int_0^\infty P(\nu,E)N(E)\,dE$$

by averaging the emission spectrum $P(\nu,E)$ of a single electron of energy E over the distribution function N(E) of relativistic electrons. Observations show that the distribution function of cosmic electrons can be approximated by a power law

$$N(E) = K E^{-\gamma} \text{ with } \gamma \sim 2.6 \qquad \text{(III.15)}$$

One can show that, for $H \simeq 3 \times 10^{-6}$ Oe, electrons with energies 10^2 MeV $\leq E$ 5×10^4 MeV are primarily responsible for the observed diffuse non-thermal galactic radiation in the frequency range 1 MHz $\leq \nu \leq$ 4 GHz. The radiation intensity I_ν is obtained by integrating the emission coefficient along the line of sight

$$\left[\frac{I(\nu)}{\text{erg cm}^{-2}\,\text{sec}^{-1}\,\text{ster}^{-1}\,\text{Hz}^{-1}}\right] = \int_0^L \varepsilon_1(\nu)\,d\ell =$$

$$= \frac{2kT_b}{\lambda^2} = \begin{cases} a(\gamma) \times 0.04\,(2.1 \times 10^8 \lambda)^{\frac{1}{2}(\gamma-1)}\, K H_\perp^{\frac{1}{2}(\gamma+1)}\, L & \text{(III.16a)} \\ b(\gamma) \times 0.04\,(2.1 \times 10^8)^{\frac{1}{2}(\gamma-1)}\, K H^{\frac{1}{2}(\gamma+1)}\, L & \text{(III.16b)} \end{cases}$$

+) Note that in empty space the numerical value of the magnetic field expressed in Oersted or Gauss is nearly the same.

Here, the wavelength λ is expressed in cm, the magnetic field H in Oe, K (as defined in eq.(III.15)) in $(erg)^{\gamma-1}$ cm^{-3}, and the line of sight L in Kpc. To express $I(\nu)$ in W $m^{-2}ster^{-1}Hz^{-1}$, one must multiply the right hand side of the equation by 10^{-3}. Eq. (III.16a) pertains to the case where all electrons see the same component $H_\perp$ in a homogeneous magnetic field. Eq. (III.16b) pertains to the case where the distributions of both magnetic fields and electron motions are isotropic. The functions $a(\gamma)$ and $b(\gamma)$ are tabulated in Table 2. For $\gamma = 2.6$, we expect a temperature spectrum of $T_b \propto \nu^{-2.8}$ for the diffuse galactic radiation in accordance with observations.

γ	0.5	1.5	2.0	2.5	3.0	4.0	5.0
$a(\gamma)$	1.65	0.198	0.143	0.120	0.111	0.116	0.147
$b(\gamma)$	1.36	0.148	0.103	0.0829	0.0740	0.0724	0.0866
$c(\gamma)$	3.94	2.35	2.09	1.97	1.94	2.08	2.50

Table 2: The function $a(\gamma)$, $b(\gamma)$ and $c(\gamma)$ used in the theory of synchrotron radiation.

The optical depth of synchrotron reabsorption (=absorption of radio emission by magnetic damping) is

$$\tau_{syn} = \int_0^L K(\nu)\, d\ell = c(\gamma) \times 2.1 \times 10^{18} (3.5 \times 10^9)^\gamma \times K H_\perp^{\frac{1}{2}(\gamma+2)} \nu^{-\frac{1}{2}(\gamma+4)} L$$

The function $c(\gamma)$ is tabulated in Table 2; ν is in Hz and L in Kpc. We see that for conditions expected in interstellar space, i.e. $H \simeq 3 \times 10^{-6}$ Oe, $K \simeq 3.3 \times 10^{-17} erg^{1.6}$ (electrons cm^{-3}), $\tau_{syn} = 1.5 \times 10^{13}\, \nu^{-2.3}\, L \ll 1$ even for extreme values of $\nu = 1$ MHz and L = 10 Kpc. Therefore, absorption by synchrotron emission in interstellar space is negligible.

(c) Free-free radiation

Free-free radiation is emitted by thermal electrons decelerated in the Coulomb-field of ions. In computing the optical depth of free-free continuum radiation, one first computes the emission spectrum $P(v_o,b)$ of a single electron with initial velocity v_o and impact parameter b. The number of collisions with impact parameters between b and $b + \Delta b$ is $N_i\ v_o\ 2\pi\, b db$, with N_i the ion density. The emission spectrum, averaged over the appropriate range of impact parameters b,

$$<P> = 2\pi\ N_i\ v_o \int_0^{b_m} P(v_o, b)\, b db \propto N_i$$

is proportional to the ion density. As $b_m \to \infty$ the integral diverges; $b_m \simeq$ Debye radius is the appropriate upper integration limit.

Next, one averages over a Maxwellian velocity distribution of the free electrons

$$N_e\ f(v_o)\, dv_o = 4\pi\ N_e\,(m/2\pi k T_e)^{3/2} \times \exp\left\{-m v_o^2/2kT_e\right\} v_o^2\, dv_o$$

and obtains the emission coefficient

$$4\pi\ \varepsilon_\omega\ d\omega = N_e \int_0^\infty <P>_b\ f(v_o)\, dv_o \propto \frac{N_e N_i}{T_e^{1/2}}$$

The absorption coefficient $K_\nu = \varepsilon_\nu / B_\nu \propto N_e\, N_i\, T_e^{-3/2} \nu^{-2}$ and the optical depth

$$\tau_c(\nu) = \int K_\nu\ d\ell \propto \nu^{-2}\ T_e^{-3/2} \int N_e^2\ d\ell$$

Here, $N_e = N_i$ has been assumed. The electron density integrated along the line of sight yields the emission measure

$$\left[\frac{E}{\mathrm{pc\ cm^{-6}}}\right] = \int_0^L N_e^{\,2}\ d\ell \qquad \text{(III.18)}$$

Quantitatively, one obtains, for the optical depth of the free-free continuum,

$$\tau_c = 3.014 \times 10^{-2} \left[\frac{T_e}{°K}\right]^{-1.5} \left[\frac{\nu}{GHz}\right]^{-2} \left[\frac{E}{pc\ cm^{-6}}\right] <g_{ff}> \qquad (III.19).$$

The Gaunt-factor

$$<g_{ff}> = \begin{cases} \{\ln 4.955 \times 10^{-2} \left[\frac{\nu}{GHz}\right]^{-1} + 1.5 \ln \left[\frac{T_e}{°K}\right]\} & (III.20a) \\ \simeq 1 & (III.20b) \end{cases}$$

Eq. (III.20a) holds for $\left[\frac{T_e}{°K}\right]^{1.5} \left[\frac{\nu}{MHz}\right]^{-1} >> 1$, eq. (III.20b) holds for $T_e^{1.5}\nu \leq 1$. Eqs. (III.19) with (III.20a) can be expressed as

$$\tau_c = 8.235 \times 10^{-2}\ a(\nu, T_e)\ T_e^{-1.35} \nu^{-2.1}\ E \qquad (III.21)$$

with $a(\nu,T_e)$ a slowly varying function, tabulated in Mezger (1967). For most purposes $a \simeq 1$. The brightness temperature is

$$T_b = T_e (1 - e^{-\tau_c}) = \begin{cases} T_e & \text{for } \tau_c >> 1 \\ \tau_c T_e & \text{for } \tau_c << 1 \end{cases} \qquad (III.22)$$

and the flux density becomes

$$S_\nu = \frac{2k}{c^2} \nu^2 T_b \Omega \propto \begin{cases} \nu^2 & \text{for } \tau_c >> 1 \\ \nu^{-0.1} & \text{for } \tau_c << 1 \end{cases} \qquad (III.23$$

(d) Radio recombination lines

After having lost, on the average, about 50 % of its original kinetic energy, a free electron will recombine with an ion. Recombinations take place to any energy level, characterized by its principal quantum number n. The recombined electron will usually reach the ground state n = 1 by radiative transitions before being again ionized. Cascade transitions $n + 1 \rightarrow n$ are the most probable. For $n \geq 50$, these transitions produce lines in the radio frequency range. The radio recombination spectrum of any ion is hydrogenic; the transition frequency is given by

$$\nu_L(n+\Delta n \rightarrow n) = \frac{R_\infty}{1+m_e/m_A} Z^2 \left[\frac{1}{n^2} - \frac{1}{(n+\Delta n)^2}\right] \qquad \text{(III.24)}$$

$$\simeq R_\infty (1+m_e/m_A)^{-1} Z^2 \frac{2\Delta n}{n^3} \left[1 - \frac{3\Delta n}{2}\right]$$

where $R_\infty = 3.289\ 847 \times 10^{15}$ Hz is the Rydberg constant for an emitter with infinite mass, m_e/m_A is the ratio of the mass of an electron to the mass of the emitting atom, and Z is the nuclear charge. To denote a certain transition, one uses the chemical symbol, the principal quantum number of the lower energy level and α for $\Delta n = 1$, β for $\Delta n = 2$ etc. Thus, the H137β-line is the result of the transition $n+\Delta n = 139 \rightarrow n = 137$ in an H-atom; the $He^+173\alpha$-line corresponds to the transition $n+1 = 174 \rightarrow n = 173$ in an He^+-ion.

To calculate the optical depth of a recombination line, we start from eq (III.7). The number N_n of recombined atoms with an electron in the lower state n of the transition is determined by the Saha-Boltzmann equation.

$$N_n = \frac{N_e^2 h^3 n^2}{(2\pi m_e kT_e)^{3/2}} \exp\left\{X/n^2 kT_e\right\} \qquad \text{(III.25)}$$

We put

$$\frac{h\nu}{c} B_{n,n+\Delta n} = \frac{\pi e^2}{m_e c} f_{n,n+\Delta n}$$

with the oscillator strength

$$f_{n,n+\Delta n} = n\, M(\Delta n)\ (1 + \frac{3}{2}\frac{\Delta n}{n} + \ldots\ldots) \qquad \text{(III.26)}$$

The function $M(\Delta n)$ is tabulated in Brocklehurst (1972) Substitution of eqs. (III.24) to (III.26) in eq.(III.7) and integration along the line of sight yields an expression analogous to eq. (III.11) for the optical depth of radio recombination lines; putting the exponential in eq. (III.25) equal to unity

$$\tau_\nu^L(TE) = 1.071 \times 10^4 \{\Delta n\, M(\Delta n)\, Z^2 (1+m_e/m_A)^{-1}\} \times \left[\frac{T_e}{°K}\right]^{-2.5} \left[\frac{E}{pc\ cm^{-6}}\right] \left[\frac{\phi_\nu}{kHz^{-1}}\right] \qquad (III.27)$$

Values of $\Delta n\ M(\Delta n)$ are given in Table 3.

Δn	$\Delta n\ M(\Delta n)$
1	0.1908
2	0.05266
3	0.02432
4	0.01397
5	0.00906

Table 3: Values of $\Delta n\ M(\Delta n)$.

In particular, for α-transitions ($\Delta n = 1$), we obtain, for the optical depth in the center of the line profile (i.e. for ϕ_ν ($\Delta\nu = 0$)),

$$\tau^L(TE) = 1.92 \times 10^3 \left[\frac{T_e}{°K}\right]^{-2.5} \left[\frac{E}{pc \cdot cm^{-6}}\right] \left[\frac{\Delta\nu_L}{kHz}\right]^{-1} \qquad (III.28)$$

Combination of eqs. (III.21) and (III.28) yields the line-to-continuum ratio

$$\left[\frac{\Delta\nu_L}{kHz}\right] \frac{T_L}{T_c} = \frac{2.33 \times 10^4}{a(\nu, T_e)} \quad \frac{1}{1+N(He^+)/N(H^+)} \left[\frac{\nu_L}{GHz}\right]^{2.1} \left[\frac{T_e}{°K}\right]^{-1.15} \qquad (III.29)$$

with $T_L = T_e \tau^L(TE)$ and $T_c = T_e \tau_c$. The factor $[1+N(He^+)/N(H^+)]^{-1}$ takes into account that ionized He^+ contributes to the free-free continuum. This equation is valid for τ_c, τ^L (TE) << 1, TE emission and pure Doppler broadening of the recombination lines. A more general solution of the equation of transfer, that takes into account non-TE effects, opacity of the continuum radiation and collision broadening, is given in Brocklehurst (1972).

Deviations from TE emission are the rule rather than the exception in the formation of radio recombination lines. The Saha-Boltzmann eq. (III.25) pertains to TE at the kinetic temperature T_e of the free electrons. After ionization, the electron has a kinetic energy determined by the excess energy of the ionizing photons. However, a

Maxwellian distribution of the free electrons is established by collisions in a very short time. The electron temperature is determined as an equilibrium between heating of the electrons by ionization and cooling by collisional excitation of forbidden transitions. Although, the effective temperature of ionizing stars ranges from ~ 3 to 5×10^4 °K, the electron temperatures in HII regions appear to be always close to 10^4 °K. Deviations of the population of the level n from its TE value are described by the b_n factor, i.e.

$$N_n = b_n \, N(TE) \qquad \text{(III.30)}$$

Although, for $n \geq 50$, b_n-values are close to unity, line amplification by stimulated emission can occur. The correction factor for stimulated emission in eq. (III.7) takes, for radio recombination lines, the form

$$\left\{1 - \frac{g_\ell n_u}{g_u n_\ell}\right\} = \left\{1 - \frac{b_{n+\Delta n}}{b_n} e^{-\frac{h\nu}{kT_e}}\right\} \simeq b_n \left\{1 - \frac{kT_e}{h\nu} \frac{d(\ell n\, b_n)}{dn} \Delta n\right\} \frac{h\nu}{kT_e}$$

and, thus, even a small over-population of the upper level $n + \Delta n$ will render the bracket, and consequently the optical depth of the recombination line, negative.

The function

$$\beta_{n,n+\Delta n}(T_e, N_e) = \left\{1 - \frac{kT_e}{h\nu} \frac{d(\ell n\, b_n)}{dn} \Delta n\right\} \qquad \text{(III.31)}$$

has been computed for the pertinent ranges of electron temperature and density (Brocklehurst, 1970). Since $\nu_L(n, n+\Delta n) \propto \Delta n$ (eq. III.24), β is independent of Δn and depends only on the principal quantum number n of the lower level involved in a transition.

The frequency of a transition $n_{\Delta n} = n + \Delta n \rightarrow n$ is, according to eq. (III.24), $\sim \Delta n$ times higher than the corresponding α-transition. For an n_α - and an $n_{\Delta n}$-transition to coincide approximately in frequency, the condition

$$n_{\Delta n} \simeq (n_\alpha + 1) \sqrt[3]{\Delta n} - \Delta n \qquad \text{(III.32)}$$

must be fulfilled. For example, the H109α- and the H137β-line are separated by only a few MHz. For two adjacent α- and β-lines, the ratio of integrated line profiles becomes, in the case of TE emission, according to eq. (III.26)

$$\frac{\int_0^\infty \tau_\nu^\beta \, d\nu}{\int_0^\infty \tau_\nu^\alpha \, d\nu} = \frac{2M(2)}{M(1)} = 0.276 \qquad \text{(III.33)}$$

Since the enhancement factor eq. (III.31) usually decreases with increasing n, non-TE effects in the β-lines are smaller and the observed ratio of integrated H137β/H109α-profiles is smaller than 0.28.

The frequency of the recombination line of $He^+ n\alpha$ is, according to (III.24), $Z^2 = 4$ times higher than that of the Henα-line. A $He^+ n'\alpha$-line and a Henα-line coincide approximately in frequency if

$$n' \simeq Z^{2/3} \, n \qquad \text{(III.34)}$$

For example, the He109α and the $He^+173\alpha$-lines are separated by a few 10 MHz. However, the $He^+173\alpha$-line is $Z^2 = 4$ times stronger than the He109α-line if their emission measures $\int N_e \, N(He^+) ds$ and $\int N_e \, N(He^{++}) ds$ are the same.

REFERENCES

Brocklehurst, M., MNRAS, 148, 417, 1970.

Brocklehurst, M. and Seaton, M.J., MNRAS, 157, 179, 1972.

Mezger, P.G. and Henderson, A.P., Ap.J., 147, 471, 1967.

VISIBLE AND ULTRAVIOLET OBSERVATIONS OF THE INTERSTELLAR MEDIUM

G.R. Carruthers

Naval Research Laboratory
Washington, USA

I. ABSORPTION SPECTROSCOPY OF THE INTERSTELLAR GAS

1. Introduction

Absorption lines in the visible wavelength range, due to the interstellar gas, have been known since 1904 when Hartmann noticed that the CaII K-line in the spectrum of the binary star δ Orionis did not share in the Doppler shift due to the binary motion observed in the other spectral lines. In spite of a number of following observations of such "stationary lines" in the spectra of other stars, however, it was not until about the 1930's that the concept of a general interstellar gas gained acceptance.

In the optical and UV spectral regions, the observed absorption lines are resonance lines, i.e. lines arising by absorption from the ground states of atoms, molecules, and ions.

The interstellar absorption lines in the optical wavelength range accessible from the ground are shown in Table I. It is seen that the number of species represented is relatively small. This list has been essentially unchanged since the work of Adams in the 1940's. This does not reflect that these species are the only ones present in the interstellar medium, but merely that these are the only abundant ones whose resonance lines fall in the ground-accessible wavelength range. Most of the common species in the interstellar gas have their

Atomic Lines			Molecular Lines			Diffuse Lines	
Atom or Ion	λ (A)	Transitions	Molecule	λ (A)	Transitions	λ (A)	Int.
Na I..........	3302.34	$3^2S_{1/2}-4^2P^0_{3/2}$	CH	4300.31	$A^2\Delta \leftarrow X^2\Pi$ (0,0) $R_2(1)$	4430.6	10
	2.94	$-4^2P^0_{1/2}$		3890.23	$B^2\Sigma^- \leftarrow$.......... $^PQ_{12}(1)$	4760.	....
	5889.95	$-3^2P^0_{3/2}$		86.39	.... $Q_2(1)+{}^QR_{12}(1)$	5780.5	3
	95.92	$-3^2P^0_{1/2}$		78.77	.... $R_2(1)$	5797.1	1
				3146.01	$C^2\Sigma^+ \leftarrow$.......... $^PQ_{12}(1)$	6203.0	1
K I..........	7664.91	$4^2S_{1/2}-4^2P^0_{3/2}$		43.15	.... $Q_2(1)+{}^QR_{12}(1)$	6270.0	1
	98.98	$-4^2P^0_{1/2}$		37.53	.... $R_2(1)$	6283.9	6
						6613.9	2
Ca I..........	4226.73	$4^1S_0 - 4^1P^0_1$	CN	3874.61	$B^2\Sigma^+ \leftarrow X^2\Sigma^+$ (0,0) R(0)		
				75.77	.... P(1)		
Ca II..........	3933.66	$4^2S_{1/2}-4^2P^0_{3/2}$		74.00	.... R(1)		
	68.47	$-4^2P^0_{1/2}$					
			CH^+	4232.58	$A^1\Pi \leftarrow X^1\Sigma^+$ (0,0) R(0)		
Ti II..........	3072.97	$a^4F_{3/2}-2^4D^0_{1/2}$		3957.74	................(1,0) R(0)		
	3229.19	$-2^4F^0_{5/2}$		3745.33	................(2,0) R(0)		
	41.98	$-2^4F^0_{3/2}$					
	83.76	$-2^4G^0_{5/2}$					
Fe I..........	3719.94	$a^5D_4 - 2^5F^0_5$					
	3859.91	$-2^5D^0_4$					

Table I: Interstellar absorption features (Munch, 1968, p.368).

lowest resonance lines at wavelengths below the atmospheric cutoff of about 3000 Å, since the resonance transitions (involving ground state) generally involve relatively large energy changes, compared to subordinate transitions (between excited states). We will see later that even those elements which are represented by moderately strong lines in the ground-based optical range (NaI, CaII) represent only a small fraction of the total Na and Ca abundance in the ISM. Na and Ca are very minor constituents in the cosmic abundance list, ranking far below such elements as C, N, O, Ne which in turn are far below H and He (Table II). Nevertheless, the gronnd-

Element	H	He	C	N	O	Ne	Na
Relative no. of atoms	10^6	10^5	400	110	890	500	2.0
Element	Mg	Al	Si	S	Ar	Ca	Fe*
Relative no. of atoms	25	1.7	32	22	7.8	1.6	3.7

*More recent determinations give Fe abundances up to 10 times this value.

Table II: Relative cosmic abundances of the elements (Spitzer, 1968).

based results are discussed in some detail because ultraviolet results have been available only a very few years. Hence, these are not nearly as complete and detailed as the visual results.

2. Theory

Consider an atom having only two energy levels, i.e. a ground state and an excited state. The energy absorption in a spectral line corresponding to the transition between these two levels is given by

u: $B_{lu}I_\nu$ (upward), $A_{ul}+B_{ul}I\nu$ (downward); l

$$\int \kappa_\nu d\nu = \int n_l s_\nu d\nu = \frac{h\nu_{lu}}{c} (n_l B_{lu} - n_u B_{ul})$$

The last term on the right-hand side can be neglected for visible and ultraviolet transitions in the interstellar gas, since for these $B_{ul} I_\nu \ll A_{ul}$. Here s_ν = atomic absorption coefficient, and $s = \int s_\nu d\nu$ = integrated atomic cross-section. Therefore

$$s = \frac{h\nu_{lu}}{c} B_{lu}.$$

Using relations between the Einstein coefficients and the oscillator strength f,

$$\overbrace{B_{lu}}^{\text{abs.}} = \overbrace{A_{ul} \frac{g_u}{g_l} \frac{c^3}{8\pi h\nu^2_{lu}}}^{\text{emm.}}$$

and

$$f_{lu} = A_{ul} \frac{g_u}{g_l} \frac{mc^3}{8\pi^2 \nu^2_{lu} c^2}$$

We find the integrated atomic cross-section in terms of the absorption oscillator strength, f_{lu}:

$$s = \frac{\pi e^2}{mc} f_{lu}.$$

The line profile is expressed in the form,

$$s_\nu = s\phi\ (\Delta\nu)$$

where s = absorption coefficient at line center.
$\phi(\Delta\nu)$ = profile function $\int(\phi(\nu)\ d\nu = 1)$
$\Delta\nu = (\nu-\nu_o)$ where $\nu_o = \nu_{ul}$ = center frequency of line.

The optical depth is given by,

$\tau_\nu = N_1 s_\nu = N_1 s\phi\ (\Delta\nu)$ where $N_1 = n_1 L$ = column density.

Two types of profile functions are of interest,

(a) Doppler broadening. For thermal motions, averaging over a Maxwellian distribution, the "Doppler width"

$$\Delta\nu_D = (\nu' - \nu_o) = \frac{\nu_o}{c}\sqrt{\frac{2kT}{M}} = b_\nu$$

where M = mass of radiating atom or ion, and the profile function,

$$\phi_{\nu_D} = \frac{1}{b\sqrt{\pi}} e^{-(\Delta\nu/b)^2}$$

$\Delta\nu$ = the distance from line center at which $\phi_{\nu D}$ is being evaluated.

<u>(b) Natural broadening (radiation damping)</u>. This results from the Heisenberg uncertainty principle, $\Delta E \Delta t \sim \hbar$ and the radiative lifetime of the upper state which is inversely proportional to the total of all transition probabilities: Quantum mechanical damping constant

$$\gamma = \sum_l A_{ul} = \frac{1}{T_{ul}}$$

$$\phi\nu_N(\Delta\nu) = \frac{\gamma}{4\pi^2(\Delta\nu)^2 + (\gamma/2)^2} \cdot$$

The factor $(\gamma/2)^2$ can be neglected for large $\Delta\nu$ in which the natural broadening is dominant over Doppler broadening, i.e. the point at which

$$\phi_{\nu_D}(\Delta\nu) = \phi_{\nu_N}(\Delta\nu) \simeq 3\ b \qquad \text{(3 Doppler widths from line center)}$$

Therefore radiation damping predominates the profile function only for very strong lines ($\Delta\nu$ more than 3 Doppler widths at the $\tau = 1$ points).

<u>Equivalent width</u> of an absorption line is the integrated absorption over all wavelengths:

$$W = \int_0^\infty \frac{I_o - I_\nu}{I_o} d\nu = \int_0^\infty \left(1 - \frac{I_\nu}{I_o}\right) d\nu = \int_0^\infty \left(1 - e^{-\tau_\nu}\right) d\nu$$

and is the width of a "square well" line (zero intensity within) having the same total absorption as the actual profile.

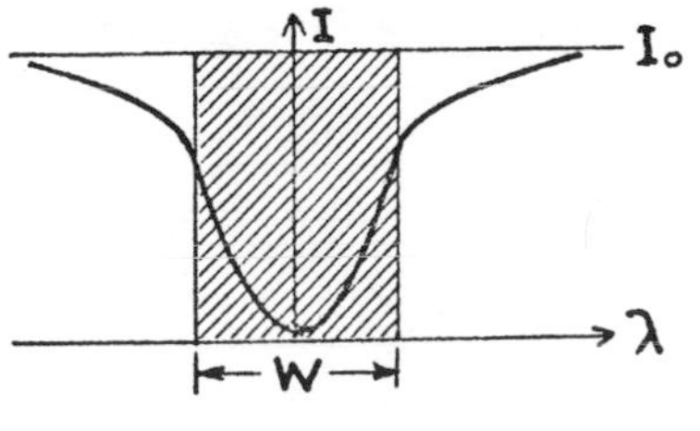

Curve of growth is the general relationship between the equivalent width of a line and the number of absorbing atoms or molecules producing it.

Consider 3 cases:

(a) $\tau \ll 1$ all frequencies
(b) $\tau \gg 1$ at line center but $\ll 1$ at $\Delta\nu = 3b$
(c) $\tau \gg 1$ beyond $\Delta\nu = 3b$.

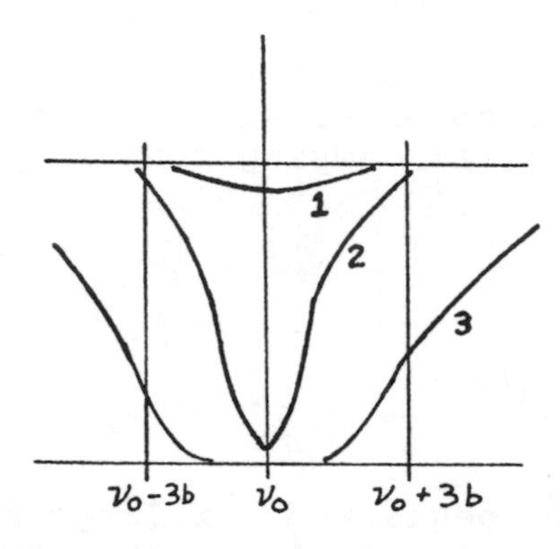

Case 1 weak line: $\tau \ll 1$ everywhere,
in this case,

$$e^{-\tau} \simeq 1 - \tau \text{ and } W \simeq \int \tau_\nu d\nu = N_1 s \int \phi_D(\Delta\nu) d\nu = N_1 S$$

therefore $W_\nu = N_1 \frac{\pi e^2}{mc} f_{lu}$ (linear region of the curve of growth).

In wavelength units,

$$W_\lambda = \frac{\lambda_0^2}{c} W_\nu \rightarrow \quad W_\lambda = N_1 \frac{\pi e^2}{mc^2} \lambda_0^2 f.$$

Case 2 intermediate strength lines: Combining the Doppler profile function with expressions for τ and W gives

$$W = 2b\, F(\tau_0) \text{ where } F(\tau_0) \equiv \int (1 - e^{-\tau_0 \exp(-x^2)}) dx$$

$$\tau_0 = \text{optical depth at line center} = \frac{N_1 s}{\sqrt{\pi}\, b} = \frac{\sqrt{\pi} e^2}{mc\, b} f_{lu}$$

For large τ_0 ,

$$F(\tau_0) = \sqrt{\ln \tau_0}$$ ("saturation" or "flat" region of curve of growth) - relatively insensitive to N_1.

Case 3 <u>very strong lines:</u>

$$\varphi(\Delta\nu) = \phi_N(\Delta\nu) = \frac{\gamma}{4\pi^2(\Delta\nu)^2} \quad \text{where } \gamma = \sum_1 A_{ul}$$

Expressing A_{ul} in terms of f_{lu}, therefore

$$\tau_\nu = N_1 s \phi_N(\Delta\nu) = \frac{N_1 s\, \gamma}{4\pi^2(\Delta\nu)^2}$$

therefore,

$$W = \sqrt{\frac{N_1 s \gamma}{\pi}}$$ (square root portion of curve of growth).

Consider the special case of a two-level atom. Expressing $\gamma = A_{ul}$ in terms of the oscillator strength f_{lu},

$$\tau_\nu = 2\pi \frac{g_l}{g_u} \left[\frac{\nu_o}{\Delta\nu} \frac{e^2 f_{lu}}{mc^2}\right]^2 N_1$$ or, in terms of wavelength,

$$\tau_\lambda = 2\pi \frac{g_l}{g_u} \left[\frac{\lambda_o}{\Delta\lambda} \frac{e^2 f_{lu}}{mc^2}\right]^2 N_1$$

$$W_\nu = \int_0^\infty (1 - e^{-\tau_\nu})\, d\nu = \frac{2\pi e^2}{mc^2} \nu_o f_{lu} \sqrt{\frac{2g_l}{g_u} N_1}$$

$$W_\lambda = \frac{2\pi e^2}{MC^2} \lambda_o f_{lu} \sqrt{\frac{2g_l}{g_u} N_1}$$

In the case where the upper state can radiate to several lower states, a somewhat more general expression is

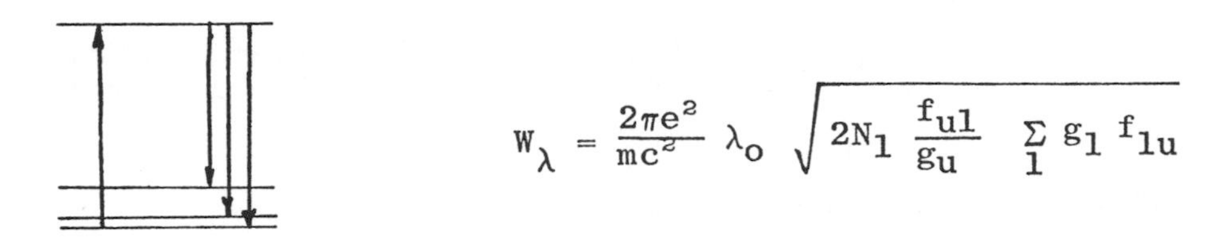

$$W_\lambda = \frac{2\pi e^2}{mc^2} \lambda_o \sqrt{2N_1 \frac{f_{ul}}{g_u} \sum_1 g_1 f_{1u}}$$

Curve of growth: This is a plot log W vs. log n (or W/b vs. log τ_o), i.e., the general relationship between the number of absorbing atoms and the equivalent width of the absorption line.

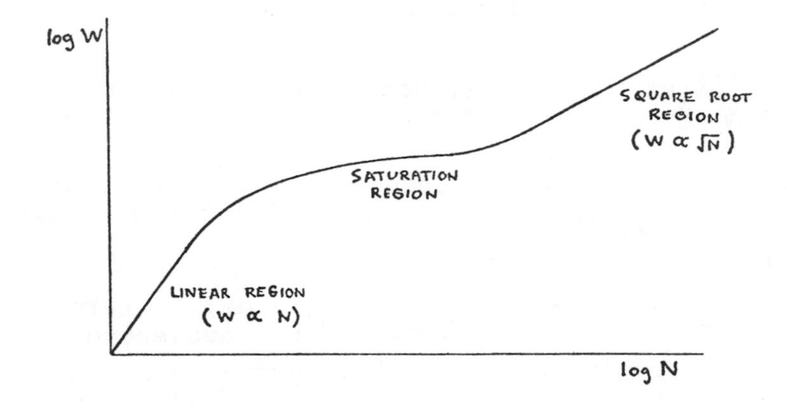

So far we have assumed the gas has no velocity component other than the thermal velocity. Actually, the gas is concentrated in discrete clouds having independent linear velocities and internal (turbulent) velocities, as well as microscopic thermal velocities. These must be taken into account in constructing the curve of growth. For example, if two clouds of equal column density are in the line of sight and have velocity separation greater than the internal Doppler width of each cloud, the profile will look like case (a) left, and the flat portion of the curve of growth will be reached for twice as large a total equivalent width (total column density) than if all the absorption were concentrated in a single line. Likewise large turbulent velocities in a cloud will require larger column density to saturate the line. For this latter case, and for overlapping line profiles, the line profile and equivalent width must be calculated for different N values to construct the curve of growth.

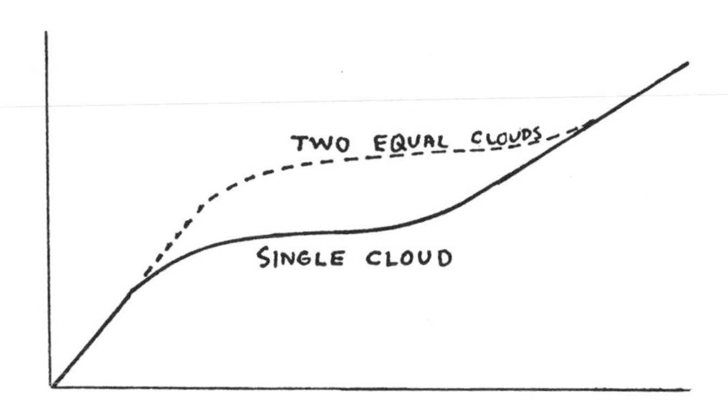

3. Ground-based observations

(a) Neutral and ionized atoms. For lines of the same element having different f-values (e.g. sodium), if the f-values are known the ratio of the line strengths can be used to determine the number of atoms more readily than from the strength of a single line, especially if one or both lines lie in the "saturated" region of the curve of growth.

The doublet ratio method, which is useful for the NaI D-lines and CaII H and K lines, makes use of the fact that the ratio of the line strengths for the two components of the doublet is the ratio of the statistical

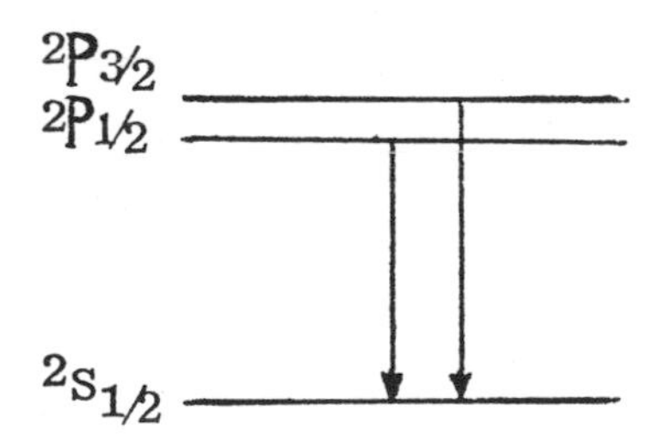

weights of the upper levels, i.e. exactly 2:1. Both levels have the same transition probability A_{ul}. For weak lines, the equivalent widths are also in the ratio 2:1, but this ratio decreases toward unity as the lines become saturated. The "doublet ratio" $F(2\tau_o)/F(\tau_o)$ is readily calculable for any value of τ_o, and is tabulated by Spitzer (1968, p.20).

For highly saturated lines, however, it is difficult to measure the doublet ratio accurately enough, as this becomes a rather weak function of the column density. For NaI, one can use the much weaker 3302-3303A doublet (f-value about 22x smaller than D-lines), and/or their relative strengths compared to the D-lines.

In observations of NaI, CaII, etc. we must consider the fact that not all the Na or Ca is in the state of ionization observed (e.g., most Na is actually NaII which is unobservable). In steady state (photoionization rate = recombination rate)

$$\frac{n_e n_i}{n_{total}} = \frac{\beta}{\alpha}$$

where β = photoionization rate coefficient

α = recombination rate coefficient

$n_{total} = n_o + n_i$

$$\beta = c\int_{\nu_i}^{\infty} \alpha_{\nu_i} U_\nu \frac{d\nu}{h\nu}$$

U_ν = radiation density

$\alpha_{\nu i}$ = atomic photoionization rate coefficient

Atom	E_i (ev)	10^2 °K	10^4 °K
CI (in HI reg)	11.3	14	320
NaI	5.1	0.68	24
CaI	6.1	230	8200
CaII	11.9	0.071	1.5
CaII (HI)		0.025	0.51

Table III: Theoretical β/α (Spitzer, 1968, p.121).

The values of β/α for NaI and CaII are relatively low, despite their low ionization potentials, because most of their oscillator strength is in the resonance doublets and relatively little is in the photoionization continuum.

Note that the NaI and CaII densities are actually higher in HII regions than HI regions, because the increased electron density (causing faster recombination rate) is much more important than the slight increase in ionization rate due to the presence (in HII regions) of ultraviolet radiation having $\lambda < 912$ Å.

In HI regions, the electron density is due to photoionization of atoms having $E_i < 13.6$ eV, and to cosmic ray ionization of H. (These are roughly comparable.) In HII regions, $n_e \simeq n_H$. Based on cosmic abundances (Spitzer, 1968, p.122),

$$\text{HII:}\quad n(\text{NaI}) = 8.3\times10^{-8}\, n_H^2$$

$$n(\text{CaII}) = 1.1\times10^{-6}\, n_H^2$$

$$\text{HI:}\quad n(\text{NaI}) = 3.8\times10^{-8}\, n_H^{3/2}$$

$$n(\text{CaII}) = 8.4\times10^{-7}\, n_H^{3/2}$$

(assuming photoionization rate ≃ cosmic ray ionization rate)

Observed: n(CaII) ≃ n(NaI) whereas n(CaII) actually should be about 30 x n(NaI); (i.e. Ca seems underabundant. The Na densities seem reasonable compared to measured values of n(HI), whereas n(CaII) implies much lower n(HI). Relative abundances seem normal in high-speed clouds. This suggests that, in clouds of normal velocity, the calcium is locked up in grains in the form of refractory compounds ($CaSiO_3$, etc.).

In dense clouds, even Na is largely condensed out. The properties of HI regions are less accurately known than HII regions, because of the uncertainties of electron density.

Herbig (1968) did a detailed analysis of the interstellar line spectrum of ζ Ophiuchi, which is relatively near and bright, but is moderately reddened (E(B-V)=0.32) and shows a rich interstellar line spectrum. The observed interstellar line spectrum results from two discrete clouds, having radial velocities of -15 and -29 km/sec; only the lines produced by the stronger (-15 km/sec) component were analysed.

It was found that the Na, K abundances relative to H were about normal, but Ca was underabundant by a factor of ~1400, and Ti was underabundant by a factor of ~100, relative to solar system values. The Ca abundance anomaly is more severe (by a factor of about 50) than observed elsewhere.

Li and Be were not observed - but the upper limits were still higher than the abundances of these elements in chondritic meteorites. Since T-Tauri stars show abnormally high Li abundances, it is of obvious interest to see if Li is in the interstellar gas from which these stars condense, or is produced in surface reactions in the T-Tauri stars during early phases of their evolution (Herbig, 1963).

In his computations, Herbig included the radiation field of ζ Oph in the ionization equilibrium equation along with the general galactic radiation field. He found that the best fit to the data resulted if the bulk of the line absorption was produced in a thin, dense cloud about 0.15 pc thick, with n_H = 500-900/cm^3.

(b) Molecules. Among the sharp interstellar lines discovered in the 1930's and early 1940's were ones due to CH, CH^+, and CN. Since then, no other molecules have been detected in the ground-based optical range; among those searched for, with negative results (although their resonance lines fall in the accessible wavelength range) are NH, OH, C_2, MgH, SiH, CO^+, N_2^+. All of these latter are observed in comets or in cool stars.

Generally, the intensities of CH and CH^+ are about equal, but the ratio can vary by a factor of 10 in either direction. The CN lines generally are much weaker.

CH, CH^+, and CN are generally believed to be produced by two-body radiative recombination (Herbig, 1963, 1970). A condition for the occurrence of two-body radiative recombination (with reasonable probability) is that the molecule must have an excited electronic state arising from ground-state atoms which can combine with a stable ground state through a permitted electronic transition: e.g. (Kaplan and Pikelner, 1970),

$$C(^3P) + H\ (^2S) \rightarrow CH\ (B^2\Sigma) \rightarrow CH\ (X^2\Pi) + h\nu$$

$$C^+(^2P) + H\ (^2S) \rightarrow CH\ (A^1\Pi) \rightarrow CH^+\ (X^1\Sigma) + h\nu$$

The potential curves involved are shown in Fig. 1. Similar reactions occur for CN and CO.

This is consistent with the absence of OH, NH, etc. and the homonuclear molecules C_2, H_2, N_2, for which this process is electric-dipole forbidden (only 3-body recombination on dust grains works for these). However, the lifetimes of the observed molecules against photodissociation by ultraviolet starlight is so short (~100 years) that it is difficult to account for the observed strengths

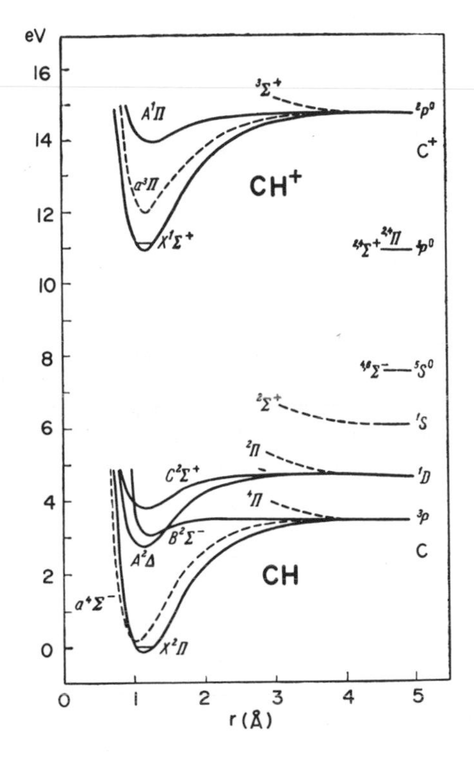

Fig. 1: Potential curves for CH and CH^+ (Kaplan and Pikelner, 1970).

of the CH and CH^+ lines by two-body recombination alone (Bates-Spitzer theory, 1951). These latter authors found a general average discrepancy of ~104 between observed and calculated densities. In the direction of ζ Oph, however, Herbig found an overabundance of (CH + CH^+) of only a factor of 30. Frisch (1972), however, showed, using newly obtained f-values for CH^+ and new rate constants by Solomon and Klemperer (1972), that the theory may account for the observations.

Another difficulty, especially in ζ Oph, is that the Bates-Spitzer theory predicts n(CH/n(CH^+) $\ll 1$, whereas the observed ratio is $\gtrsim 1$. This prediction results from the fact that CH can be photodissociated by about 10 eV photons, whereas CH^+ requires ~14 eV photons. An alternate origin of the observed molecules (Herbig, 1963, 1970) is by dissociation of CH_4 sublimated from dust grains near hot stars. This would explain abnormally strong CH^+ lines observed in the Pleiades. The lines indicate that the gas is approaching the stars. Since CH_4 sublimates at $\sim 30^oK$, vs. $\sim 80^oK$ for NH_3 and $\sim 120^oK$ for H_2O, this would explain the relative absence of NH and OH; however, CN still represents a problem.

Since there is evidence that the interstellar dust grains consist at least in part of graphite, and all of the observed molecules contain one atom of carbon, there is a suggestion that the dust grains may somehow be involved in the production of the observed molecules.

Since the radiative lifetimes of the observed molecules are short compared to mean time between collisions, in optically accessible clouds, the molecules are in equilibrium with the 3^oK black body background radiation rather than with the kinetic temperature of the gas. The excitation of low-lying rotational levels of molecules in thermal equilibrium is given by

$$\frac{n_J}{n_{total}} = \frac{g_J\, e^{-E_J/kT}}{Q}$$

where Q = rotational partition function
$E_J = B\,J(J+1)$

For CN, B is low enough that the first excited rotational level can be populated by the 3^oK black body microwave background radiation (Fig. 2).

(c) Diffuse interstellar bands. These have intrinsic half-widths up to ~ 10 Å - the feature near λ 4430 is the strongest. About 15 features known between 4400 and 6700 Å (Herbig, 1963).

Strengths of these features correlate more strongly with dust (reddening) than with atomic line strengths. The presence of hot stars reduces the strength of these bands; the absorbing properties of the dust may also be modified at the same time. The great strength of the diffuse bands indicates they must involve cosmically abundant elements.

4. Far ultraviolet observations

(a) Atomic hydrogen. Hydrogen is by far the most abundant element in the universe and in the interstellar gas. Although atomic hydrogen can be detected by its

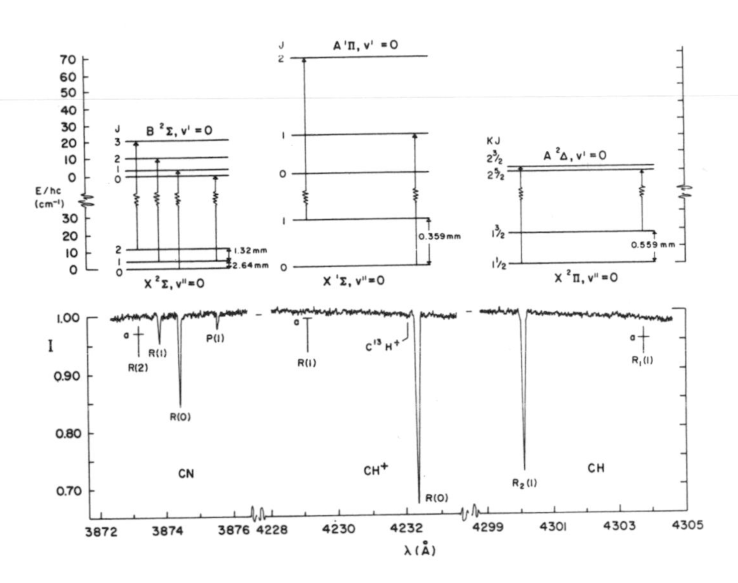

Fig. 2: Interstellar lines of CH, CH^+, and CN in ζ Ophiuchi (Bortolot et al., 1969).

emission (or, in a few rare cases, absorption) of 21-cm radiation, it has long been realized that much more accurate and detailed studies of the spatial distribution of interstellar H could be made by observing its resonance absorption lines (e.g. Lyman-α at 1216 Å) in the line of sight to hot stars (which provide continuum background light sources).

Since in almost all cases the Ly-α absorption is so strong that the profile is dominated by natural broadening, we use the expression for the square-root region of the curve of growth. Substituting numerical values for H Lyman-α(Morton, 1967; Jenkins, 1970)

$$\tau_\lambda = 4.26 \times 10^{-20} \frac{N_H}{(\lambda-\lambda_o)^2}$$

$$W_\lambda = 7.31 \times 10^{-10} \sqrt{N_H}$$

Theoretical profiles for the interstellar Lyman-α absorption for representative H column densities are shown in Fig. 3. It is seen that the interstellar hydrogen can be readily measured with relatively low spectral resolution.

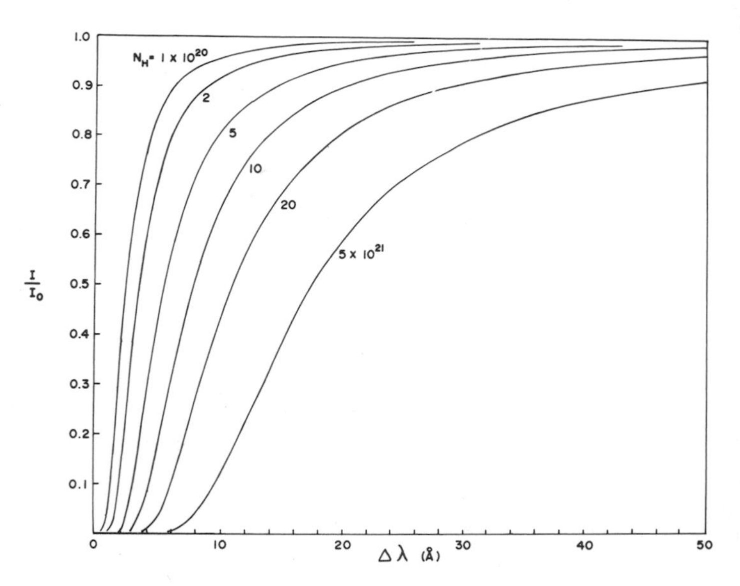

Fig. 3: Theoretical profiles of the hydrogen Lyman-α absorption for various column densities, assuming pure natural broadening.

The first observations of the interstellar Lyman-α were made in a rocket flight by Morton (1967) and co-workers at Princeton. They used objective grating Schmidt spectrographs to observe hot stars in the constellation Orion. They observed equivalent widths of the Ly-α line generally less than 10 Å, corresponding to column densities of $\sim 10^{20}/cm^2$ ($\sim 0.1/cm^3$) which was much less than 21-cm radio measurements which indicated $\sim 10^{21}/cm^2$ ($1/cm^3$) These results were confirmed by others at NRL (Carruthers, 1968, 1969) and GSFC (Smith, 1972a) (Fig. 4), in Orion, and even lower column densities were observed toward ζ Pup and γ Vel (Carruthers, 1968; Morton et al., 1969; Smith, 1970) indicating < 0.06 H atoms/cm^3. In the directions of the Perseus and Scorpius-Ophiuchus associations, however, much higher column densities were found, corresponding to 1 H atom/cm^3 (Morton et al., 1972a, 1972b).

OAO-2 greatly added to the available sky coverage. However, the spectra had only about 12 Å resolution, so early results overestimated the H column densities due to blending with near by stellar lines. However, Savage and Jenkins (1972) used higher-resolution rocket spectra to correct the OAO profiles, and H column densities were found by multiplying the observed curves by $e^{+\tau}$ (using

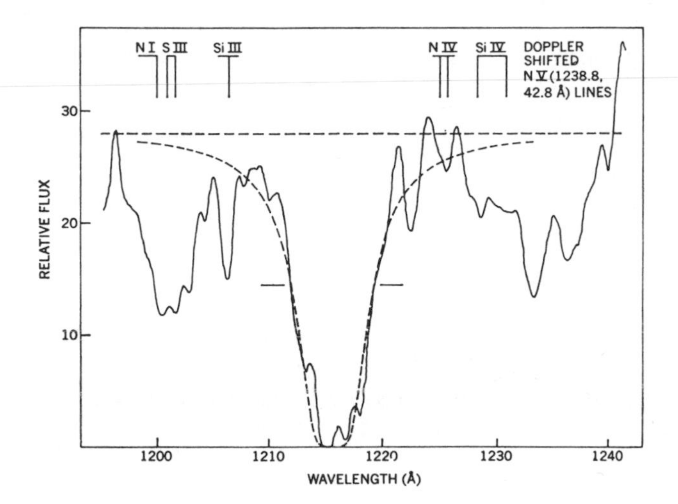

Fig. 4: Observed interstellar Lyman-α absorption in ζ Orionis, with best-fitting theoretical profile (Smith, 1972a).

computed τ values corresponding to a range of H column densities) until a flat continuum was produced. This method was felt to be more accurate than attempting to find the $e^{-\tau}$ profile best fitting the observed profile.

They found that the H density varies over a wide range (from < 0.1 to > 1.0) with the average for all the observed stars ~ 0.6 atoms/cm^3. In the near-solar neighborhood, $N_H \sim 0.25/cm^3$; also a general deficiency was found in the sector $180° < l_{II} < 270°$. The hydrogen column densities correlated fairly well with color excess, E(B-V), D-line, and H-K line measurements, yielding

$$\left\langle \frac{N_H}{E(B-V)} \right\rangle = 5 \times 10^{21} \text{ at/cm}^2 \text{ mag}$$

$$\left\langle \frac{N_{NaI}}{N_{HI}} \right\rangle = 3.5 \times 10^{-9} \rightarrow \left\langle \frac{Na}{H} \right\rangle \simeq 2.3 \times 10^{-7}$$

$$\left\langle \frac{N_{CaII}}{N_{HI}} \right\rangle = 2.5 \times 10^{-9} \rightarrow \left\langle \frac{Ca}{H} \right\rangle \simeq 6.8 \times 10^{-9}$$

From a recent grain model, they find $\left\langle \frac{\rho_{HI}}{\rho_{dust}} \right\rangle \simeq 100$.

However, the correlation with 21-cm measurements was very poor. This was felt mainly to be the result in the differences of sampling geometry between Lyman-α absorption and 21-cm emission measurements.

(b) Other neutral and ionized atoms. Intermediate in strength between the ground-accessible interstellar lines and the hydrogen Lyman lines are expected to be the interstellar resonance absorption lines of the common elements C, N, O, Mg, Si, P, S, etc. As early as 1967, Morton and co-workers (Stone and Morton, 1967) at Princeton tentatively identified some of these interstellar lines in the far-ultraviolet spectra of Scorpius stars, but the limited spectral resolution (~ 1 Å) prevented separation of interstellar lines from stellar lines at nearly the same wavelengths. However, later rocket flights by Princeton (Morton et al., 1972a) and by Smith (1972b) at GSFC attained resolutions of the order ~0.5 Å in far-ultraviolet spectra of ζ Ophiuchi, and the stars in Scorpius and Perseus (Morton et al., 1972b). Fairly definite identifications of CI, CII, OI, SiII, SII, NI, NII were made in these experiments.

Smith (1972b) did a detailed curve-of-growth analysis of his spectrum of ζ Oph, following that of Herbig (1968) in the visible range. He found that the CII, OI, SiII lines fell on the square root region of the curve of growth and the calculated abundances relative to hydrogen appeared normal. The CI and SII lines were on the flat portion of the curve of growth, and it was necessary to assume the absorbing material to be distributed among several clouds, as in the visible, to obtain a proper fit.

In addition to the 1334.5 Å line of CII arising from the lowest ($^2P_{1/2}$) fine-structure level of the ground state, an even stronger feature (1335.7 Å) arising from the excited $^2P_{3/2}$ (.0079 ev) level was observed. Smith postulated that this absorption was produced within the HII region, presumably in a circumstellar envelope resulting from mass ejection by ζ Oph (evidenced by the P-Cygni profile of the CIV resonance near 1550 Å).

Jenkins (1973), at Princeton, also observed the CII with excited component in ζ Oph, δ Sco, ζ Per. However, he found greater values of N(CII)/N(HI) (3X cosmic abundance).

OAO-3 (Copernicus). The Princeton experiment package aboard OAO-3 (Copernicus) is intended specifically for studies of interstellar absorption lines. It obtains

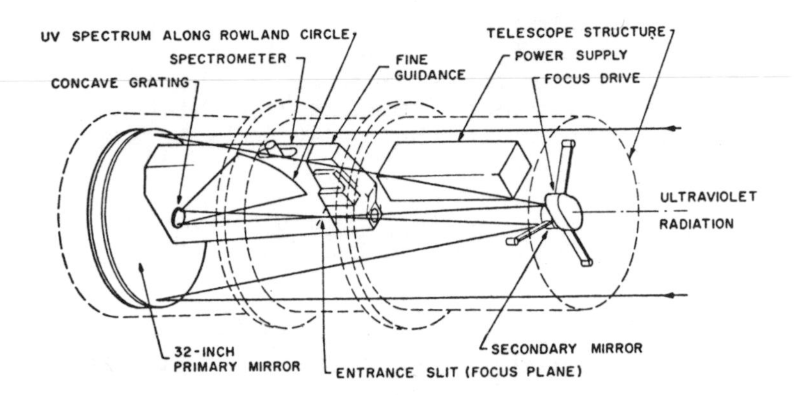

Diagram of the Princeton experiment package on board OAO-3 (Copernicus).

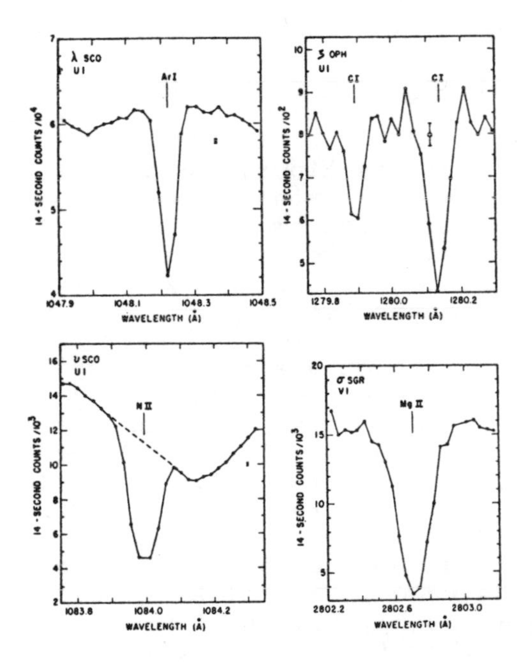

Sample spectra taken with the Princeton ultraviolet spectrometer on OAO-3 (Copernicus).

spectra of 0.05 (or 0.5) Å resolution between 950-1450 Å, and 0.1 (1.0) Å resolution between 1650 and 3000 Å. Unreddened B stars down to m_V=5.0 can be observed at 1100 Å with good signal-to-noise ratio (m_V=3.0 in the 1650 - 3000 Å range). The preliminary results (on which the following is based) appear in the 1 May 1973 Astrophysical Journal Letters.

<u>Observations of interstellar clouds</u>: Observed species in the reddened stars ζ Per, ζ Oph, α Cam, λ Ori, γ Ara include:

HI	CI	NI	OI	Mg I			S I		Ar I		
	C II	NII		Mg II	Si II	P II	S II	Cl II		M II	Fe II
					Si III		S III				
					Si IV		S IV				

Curve of growth analysis showed that the ratios of Mg, P, Cl, Mn to H were less than in the sun by factors of about 4 to 10. Several other elements seem depleted in some of the stars. Excited levels in the ground terms of CI, CII, NII, and SiII were also found present.

<u>Observations of the intercloud medium</u>: Interstellar lines were studied in the unreddened stars (E(B-V)<0.03) λ Sco, ν Sco, α Leo and α Eri, located at distances 20-150 pc from the sun. The lines were on the linear and saturated regions of the curve of growth. Preliminary analysis showed that processes producing large amounts of highly ionized species are not dominant. Doubly ionized C, N, Si, and S have been detected, but are all quite weak relative to the singly ionized species.

Abundances relative to nitrogen are in the cosmic ratios or slightly lower, with some indications of general depletion of the heavier elements. Hydrogen number densities range from a measured value of 0.22/cm^3 for λ Sco to a derived value of 0.02/cm^3 for α Leo, assuming nitrogen to have its solar abundance relative to hydrogen.

The Utrecht experiment (De Boer et al., 1972) aboard the ESRO TD-1A satellite observed interstellar lines of MgII, MgI, FeII, MnII, and possibly FeI in ζ Pup. These observations also indicated that these heavy elements, like Ca in the visible, are depleted in the interstellar gas relative to solar abundances.

<u>(c) Molecular hydrogen and carbon monoxide</u>. Molecular hydrogen has been of interest to studies of the inter-

stellar medium for many years; since hydrogen composes most of the mass of the interstellar gas, molecular hydrogen would logically be expected to be one of the major molecular constituents. Molecular hydrogen has been invoked to explain the so-called "missing mass" of the galaxy, and the apparent deficiency of atomic hydrogen 21-cm emission in dense dust clouds, where the total gas density would be expected to be greatest. Unlike atomic hydrogen, H_2 has no radio emission; it can be detected only by observation of its far-ultraviolet resonance absorptions (1108 Å and shorter wavelengths) (see Fig. 5).

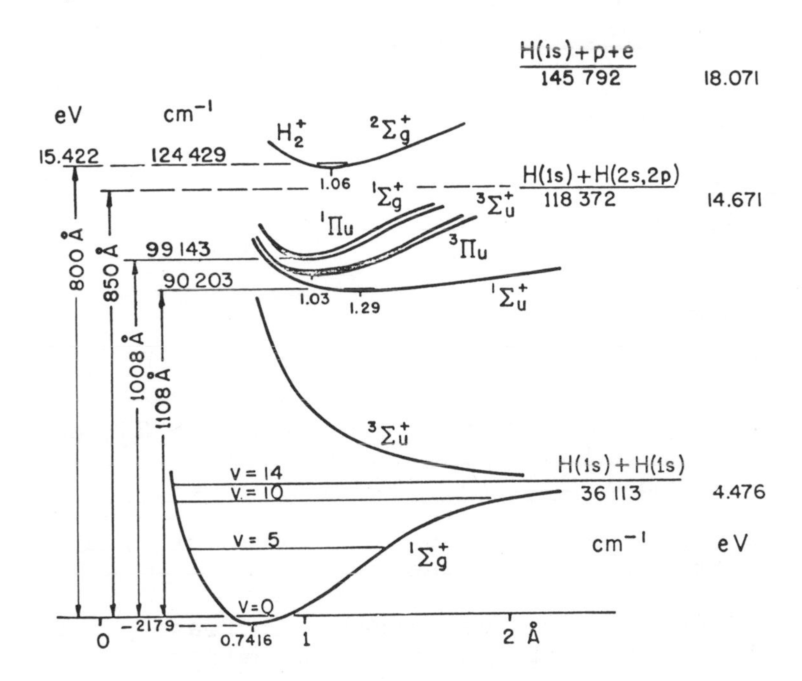

Fig. 5: Potential curves of H_2, showing primary resonance transitions and direct photodissociation and photoionization transitions (Field et al., 1966).

This figure also shows why H_2 was expected to be an abundant molecule, because it cannot be directly photodissociated by any radiation longward of the Lyman limit. Hence, some early estimates indicated that H_2 could be more abundant than H, and could possibly explain the "missing mass". Unlike CH, CN, etc., H_2 cannot be formed by two-body radiative recombination; it can be formed only by 3-body recombination on a dust grain (which takes up the energy of recombination):

$$H + H + M \rightarrow H_2^* + M \rightarrow H_2 + M^*.$$

The heat deposited in the grain by the recombination assures that the H_2 escapes, rather than forms a monolayer (which would tend to inhibit further recombinations).

However, the early estimates overlooked indirect processes of H_2 photodissociation (Fig. 6), which are so effective that in the general interstellar medium, $H_2/H < 10^{-7}$.

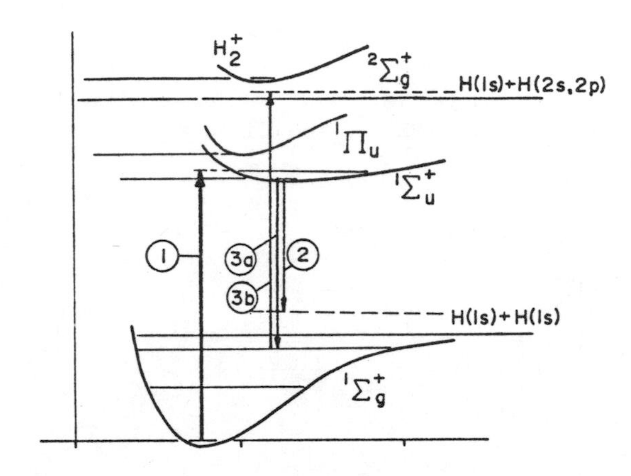

Fig. 6: Indirect photodestruction processes for H_2: absorption in the Lyman bands (1), followed by (2) or (3a) and (3b) (Stecher and Williams, 1969).

In dense dust clouds, however, H_2 could still be the dominant form of hydrogen, because

(a) the dust grain and gas density increase provides an increase in the rate of 3-body recombination,

(b) H_2 is shielded from photodissociating starlight by dust and other H_2 molecules.

The same types of calculations given previously for atomic H apply to H_2 with approximate distribution over the rotational levels of the ground state, given by the Boltzmann formula (Fig. 7). Since pure rotational transitions of H_2 are dipole forbidden, the radiative lifetime is not small compared with the time between collisions, especially in the dense clouds in which H_2 is found. Hence, the rotational excitation is more closely in equilibrium with the gas kinetic temperature than with the 3° background radiation.

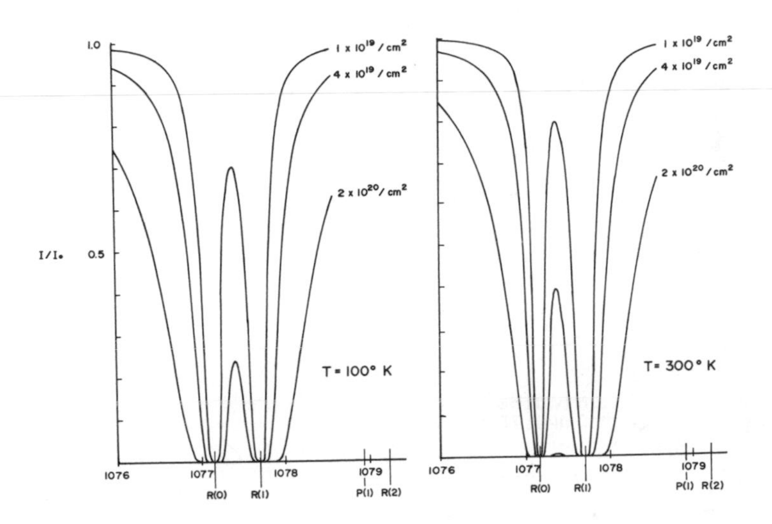

Fig. 7: Theoretical profiles of H_2 absorption in the R(0) and R(1) lines of the (2-0) Lyman bands, for temperatures of 100° and 300^oK.

Laboratory spectra (Carruthers and Bromberg, 1971) of H_2 Lyman-band absorptions have been made with an electronographic spectrograph intended for rocket spectroscopy of hot stars. At H_2 column densities comparable to the H column densities, the bands are quite strong, and are apparent at relatively low spectral resolution. At the highest column densities, the damping wings of the H_2 bands almost totally deplete the continuum below 1120 Å.

Fig. 8 shows a spectrum of the O7 star ξ Per, obtained by NRL in March, 1970. It shows absorptions of interstellar H and H_2. It was estimated that about 40% of the hydrogen was in the form of H_2.

Fig. 9 shows one of the H_2 bands in ξ Per observed with much higher resolution by OAO-3. It shows that the higher rotational levels are populated corresponding to a gas kinetic temperature of $150-200^o$ K.

In H_2, levels with odd J (orthohydrogen) cannot combine, radiatively, with levels of even J (parahydrogen). Hence, the relative populations of J=0 and J=1 reflect the orth-para formation ratio. The "temperatures" determined from this ratio are somewhat lower ($\sim 80^o$ K).

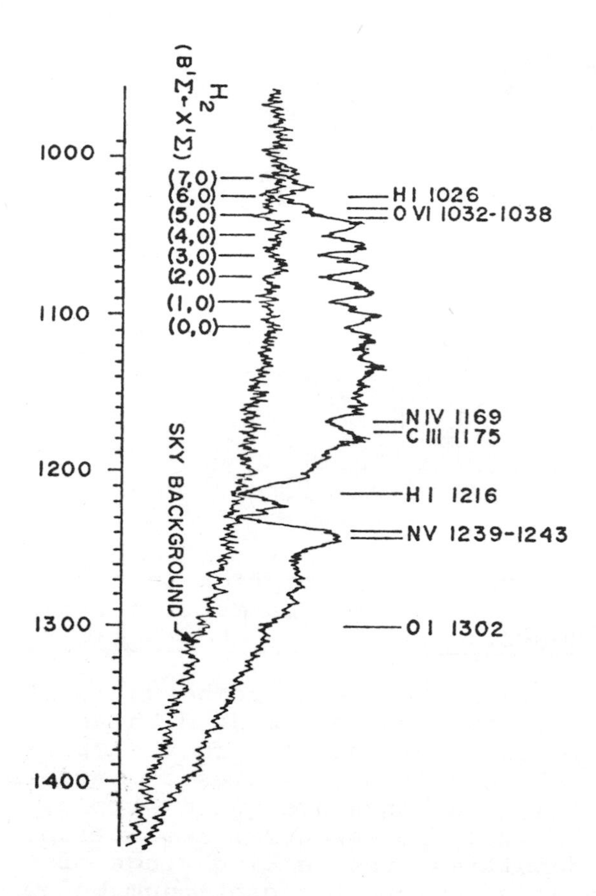

Fig. 8: Far-ultraviolet spectrum of ξ Persei, showing absorptions of interstellar atomic and molecular hydrogen (Carruthers, 1971).

For all stars thus far observed having $E(B-V) > 0.10$, the fraction of hydrogen in molecular form exceeds 0.1. However, for stars with $E(B-V) < 0.05$, there is no trace of H_2 absorption, with less than 10^{-7} of the hydrogen in molecular form. Only two intermediate cases have so far been reported, with f in the 10^{-5} - 10^{-6} range. Given below are some of these observations.

Star	E(B-V)	f
ζ Oph	0.32	0.67
ξ Per	0.32	0.46
λ Ori	0.12	0.11
ζ Ori	0.09	1.8×10^{-5}

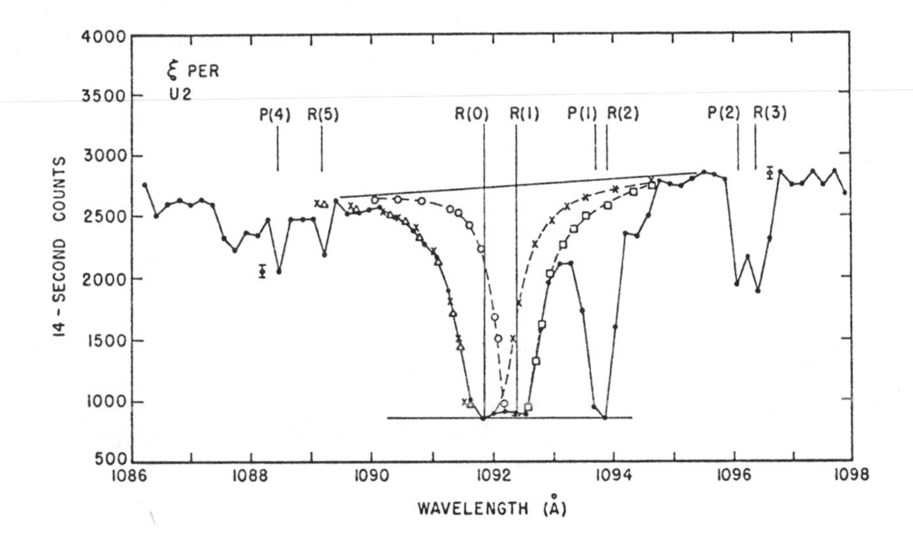

Fig. 9: Spectrum of the (1-0) Lyman band of H_2 in ξ Persei obtained by OAO-3 (Spitzer et al., 1973).

Measures of HD lines in 9 stars indicate a ratio of HD to H_2 equal to 10^{-6}. Correction for the more rapid photodissociation of HD, in the absence of optical shielding by many other HD molecules indicates one HD molecule is formed and dissociated for every 200 H_2 molecules, hence the relative abundance of D to H is about 5×10^{-3}, vs. 2×10^{-4} on the earth. Because of the uncertainties in the shielding correction and the unknown relative HD formation efficiency, however, a much lower D/H ratio cannot be ruled out.

Carbon monoxide, after H_2, is expected to be the most abundant interstellar molecule. This is due, not only to the relative abundance of its components and the fact that it, like CH, CH^+ and CN, can be formed by two-body recombination, but primarily because it is the most stable diatomic molecule known. It has a dissociation energy of > 11 eV, corresponding to photons of $\lambda < 1180$ Å. Therefore, it has a much longer lifetime against photodissociation than do CH and CN.

The first positive detection of CO was made by Smith & Stecher (1971), who observed the Fourth Positive Band System ($A^1\Pi - X^1\Sigma$) in ζ Oph between 1200 and 1600 Å. They found a column density of $\sim 7.6 \times 10^{15}/cm^2$ (100xCH), and also observed the isotopic C^{13} form of CO, with a C^{12}/C^{13} ratio $\simeq 105$.

This is in close agreement with the C^{12}/C^{13} ratio of ~ 80 for the CH^+ band in ζ Oph determined by Bortolot and Thaddeus (terrestrial ratio = 89).

The curve of growth could not be fitted unless it was assumed that the rotational levels J=0, J=1 (producing R(0), R(1), and Q(1) lines which were not resolved) were populated by black body background radiation at 3°K, giving n(J=1) = 0.47 n(J=0). They assumed this equivalent to 3 clouds separated by velocity components equivalent to the rotational line separations.

OAO-3 observed the 4th-positive bands of CO in ζ Oph, deriving a column density of $4.6 \times 10^{14}/cm^2$ (~ 10 x less than Smith and Stecher). The much stronger transitions $C^1\Sigma^+ - X^1\Sigma^+$ (1087.9 Å) and $E^1 - X^1\Sigma^+$ (1076.1 Å) were observed in high-resolution spectra of λ Ori, ξ Per, and α Cam. The column densities measured were ~ 10-20 x lower than ζ Oph. The measured rotational temperature was ~ 5° K. For a gas kinetic temperature of 20-50° K, this excitation was found consistent with a number density of neutrals of 10-100/cm^3.

II. EMISSION SPECTROSCOPY OF INTERSTELLAR GAS

The interstellar gas can be observed in emission in the visible and ultraviolet wavelength ranges, only when it is excited by the ultraviolet light of hot stars or (in some cases) by collisions with high energy particles or by shock waves.

Generally, emission spectroscopy (even when suitable excitation is present) is limited by the available experimental capabilities to gas densities considerably greater than the average in interstellar space. The brightness of emission nebulae are expressed by the emission measure

$$E = n_e^2 \, L \ \mathrm{pc\ cm^{-6}}$$

where n_e = electron density, L = pathlength through the emitting region (pc), since the emission rate is proportional to the square of the electron density (for the same electron temperature, usually ~ 10^4 °K, and no (or same) interstellar extinction, HII regions with the same abundances relative to H).

The minimum value of E for which accurate photometry is possible is ~ 100-200. In the brightest parts of the Orion nebula, however, E = 10^7.

The sizes of HII regions or "Strömgren spheres" can be predicted from the expression,

$$L_{lyc} \text{ (phot/sec, } \lambda < 912 \text{ Å)} = \frac{4}{3}\pi r_s^3 n_H^2 \alpha^{(2)}$$

where

L_{lyc} = total photon flux of the exciting star at $\lambda < 912$ Å,

r_s = radius of the Strömgren sphere, and

$\alpha^{(2)}$ = recombination rate to levels $n \geq 2$.

The following table (Spitzer, 1968, p. 177) gives properties of HII regions for various spectral classes of exciting stars.

Spect. Type	Tc	L_{Lyc}	$r_s n_H^{2/3}$
05	56,000 °K	31 x 10^{48}/sec	100 pc cm^{-2}
07	36,000	2.7	44
09	25,000	0.19	18
B1	18,000	0.008	6.4

We see that, even for an 05 star, the emission measure will be only about 100 if n_H has the slightly-above-average value of $1/cm^3$. Hence, optical measurements of HII regions preferentially select areas of considerably greater than average gas density.

Emission nebulae derive their line emission from two main sources:

(1) Recombination spectra (cascade process) - this is the main process for permitted line spectra.

(2) Electron collisional excitation - this is the process mainly responsible for the <u>forbidden lines</u>, in which radiative lifetimes are comparable to time between collisions.

Recombination spectra of the following are observed;

the spectrum of each requires photons of energy greater than the ionization potential (Dufay, 1968, p. 63):

	H	C^+	He	N^+	O^+
E_i(ev)	13.54	24.28	24.48	29.49	35.00
λ(Å)	912	510	506	420	354

	N^{++}	C^{++}	He^+	O^{++}	C^{+++}
E_i(ev)	47.24	47.67	54.17	54.71	64.22
λ(Å)	262	260	229	226	193

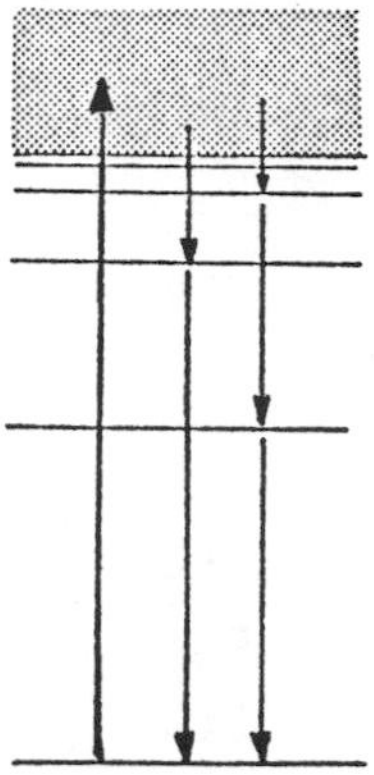

Generally, recombination spectra of species having ionization energies equal to or greater than that of He^+ are not observed in diffuse nebulae, although they are seen in planetary nebulae and other special objects. For example, the HeII recombination spectrum is not usually observed in diffuse nebulae. This probably reflects the relative lack of photons from the exciting star in the appropriate wavelength range ($\lambda < 229$ Å).

Forbidden lines are mainly excited by electron collisions. Since the forbidden lines arise from relatively low-lying metastable levels, they can be excited by relatively low-energy electrons (~5 eV). Since the radiative lifetimes, though long, are shorter than those of other processes of de-excitation, the intensities of the forbidden lines are of the same order as permitted ones of the same excitation energy ($N_f A_f \sim N_a A_a$). The electron temperature in HII regions (Pottasch, 1965) can be determined by measuring ratio of intensities of OIII lines (Fig. 10).

For electron density $< 10^3$,

$$\frac{I\ (4363)}{I(5007+4959)} = 0.137 \times 10^{-14.300/Te}$$

Unfortunately, for only a very few nebulae is the 4363 Å line strong enough to be accurately measurable.

An alternate method, which can be applied if the ratio of O/H is known, is to measure the intensities of

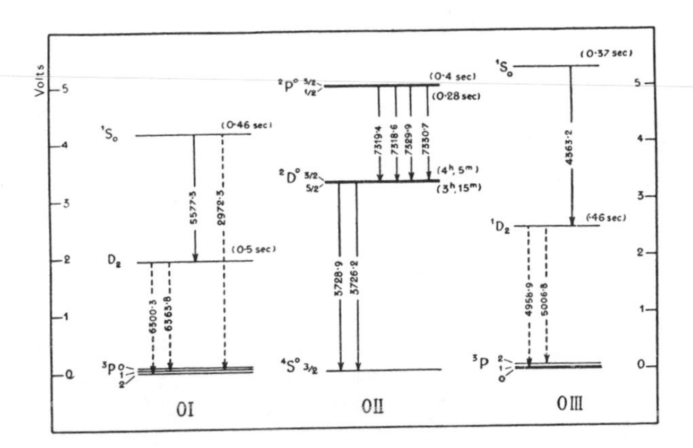

Fig. 10: Energy levels of oxygen in three lowest stages of ionization, and observed optical transitions (Dufay, 1968, p. 52).

the (OII) and (OIII) lines relative to the hydrogen Balmer lines, where

$$\frac{N(O)}{N(H)} = \frac{N(O^{+})}{N(H)} + \frac{N(O^{++})}{N(H)}$$

The forbidden lines of OII (3727 + 3729 Å) and of OIII (5007 + 4959 Å) are used.

A final method is to assume that the gas temperature and electron temperature are the same and to measure the Doppler width of the Hα line.

Continuum radiation from emission nebulae is derived from the following three sources (Pottasch, 1965):

(1) Free-bound continua (primarily of H, e.g. Balmer continuum, Paschen continuum).

$$E_{fb}\, d\nu = 2.16 \times 10^{-32}\, g\, \frac{n_i n_e}{T_e^{3/2}} \frac{1}{n^3}\, e^{-\frac{h(\nu-\nu_n)}{kT_e}}\, d\nu \quad \frac{\text{ergs}}{\text{cm}^3\,\text{sec}}$$

(2) Free-free continuum

$$E_{ff}\, d\nu = 6.72 \times 10^{-38}\, g\, \frac{n_i n_e}{T_e^{1/2}}\, e^{-\frac{h\nu}{kT_e}}\, d\nu$$

(3) Two-quantum emission (Spitzer-Greenstein)

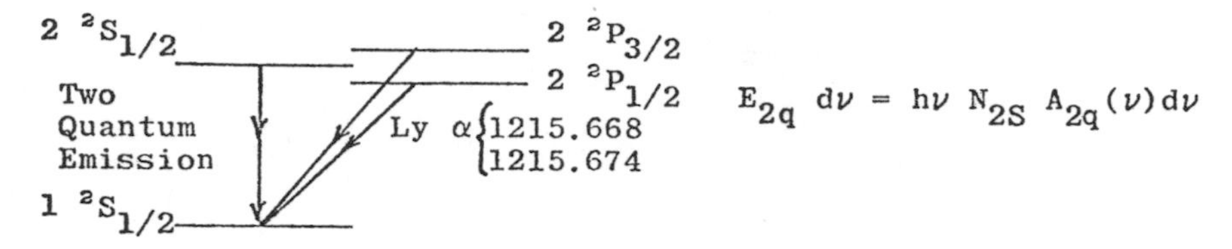

$$E_{2q}\, d\nu = h\nu\, N_{2S}\, A_{2q}(\nu) d\nu$$

where the transition probability $A_{2q}(\nu)d\nu \simeq 7 \times 10^{-15}$ in the visible spectral range (slowly varying with wavelength). Estimates can be made of N_{2S} under low-density conditions in gaseous nebulae, with the only collisional transitions of importance being the intralevel transitions ($^2S-^2P$), and with the gas optically thick in the Lyman lines. It is found that the computed intensity and spectral distribution of the 2q continuum are in reasonable agreement with observations. For T_e=10,000°K, relative contributions at λ= 5800 Å are:

2q = ≃ 8

fb = 3

ff = 0.5

Reflection of the light of the central stars by dust is found to be inadequate by a large factor (except e.g. in the outer regions of the Orion nebula) to explain the observed continua of HII regions.

The North America nebula is an example of an extended HII region. It is probably excited by several widely-separated stars and, hence, is difficult to correlate with the simple Strömgren Sphere Model.

The Cygnus loop nebula is an example of a special object, in that there is no star associated with it that could be responsible for the observed excitation. The ring shape suggests that it may be the remnant of a supernova explosion about 50,000 years ago, and the excitation may result mainly by collisions; however, since the radial velocity of the filaments is only ~ 100 km/sec, the excitation may be the result of a shock front in the surrounding interstellar gas, preceding the outward-moving supernova shell.

The NRL Far-Ultraviolet Camera/Spectrograph (Carruthers and Page, 1972) performed the first astronomical observations from the lunar surface during the Apollo 16 mission 21-23 April 1972. The experiment was based on a

75 mm aperture, f/1.0 Schmidt camera, using electronographic recording and sensitive only in the far-ultraviolet wavelength range below 1600 Å. One of the reasons for observing from the lunar surface rather than from near earth orbit, was to study faint, diffuse sources of ultraviolet light without interference from the diffuse Lyman-α glow of the hydrogen geocorona, which extends to $>$ 100,000 km from earth.

Fig. 11 is a photograph of the Cygnus region in the 1250-1600 Å range, which is believed to contain the first far-ultraviolet images of emission nebulae - the North America nebula and the Cygnus loop nebula. No spectra of these objects are available, but strong lines predicted by Osterbrock (1963) in this wavelength range from emission nebulae include CII, CIV, and SiII. In the ultraviolet, most of the predicted lines are resonance lines instead of forbidden lines; much resonance scattering thus occurs before the radiation escapes the nebula.

Also obtained was a photograph of the M8 region in Sagittarius. However, the size of this nebula is not enough greater than the resolution of our camera to allow ready separation of the nebular emission from the light of the hot stars contained within it. This was also true in the Large Magellanic Cloud; e.g. the 30 Doradus complex.

This problem, which is largely due to the fact that the exciting stars are much brighter (relative to the emission nebulosity) in the far-ultraviolet than in the visible, has also prevented measurement of the Orion nebula separately from its exciting stars, either by earlier NRL rocket flights of a similar camera or by the OAO-2 Celescope experiment. It is hoped to study the Orion region in more detail in an upcoming rocket flight.

Finally, we can obtain some information about the interstellar gas by very close-hand observations; i.e. the interaction of interstellar gas entering the solar system with solar ultraviolet radiation and solar wind particles. This yields emission lines of H Lyman-α and He 584 Å - the only direct observation of local interstellar He. The Apollo 16 experiment (Carruthers and Page, 1972) and an ultraviolet photometer aboard Pioneer 10 (Carlson, 1972) have indicated interplanetary He 584 Å intensities of about 1% of the H Lyman-α intensity.

Fig. 11: Apollo 16 photograph of a region in Cygnus (wavelength range 1250-1600 Å) showing the Cygnus loop nebula (lower right) and North America nebula (upper right). (NRL photo)

III. INTERSTELLAR DUST

1. Interstellar extinction curves

The first evidence for extinction of starlight by interstellar dust was the direct observation of obscured regions (dust clouds) in which star counts were markedly less than in other regions.

Even more convincing evidence, however, was provided by the fact that stars in such obscured regions appear redder in color than do stars of the same spectral class elsewhere. This latter fact can be used to deduce the presence of obscuration even in regions where it is not otherwise obvious.

From observations of nearby stars, known or suspected not to be appreciably reddened by interstellar dust, we can predict the spectral distribution of the continuous emission, or color, of a star of a given spectral class. We define the color excess:

$$E(B-V) = (B-V)_{observed} - (B-V)_{intrinsic}$$

Intrinsic (B-V) for various spectral classes (luminosity classes III-V) are as below (Johnson, 1968):

Spect	B-V	Spect	B-V
O5-7	-.32	B3	-.20
O8-9	-.31	B5	-.16
BO-O9.5	-.30	B7	-.12
BO.5	-.28	B9	-.06
B1	-.26	AO	.00 (by definition)
B2	-.24		

However, this color excess E(B-V) (or corresponding excesses between any other two spectral bands) only represents a differential absorption, or reddening of the starlight; it does not give the total absorption at a given wavelength.

To obtain the total extinction, one can do one of the following:

(a) Find the color excesses over a wide range of wavelengths, extending as far as possible into the infrared, and extrapolate the resulting curve to zero wavelength (assuming absorption there):

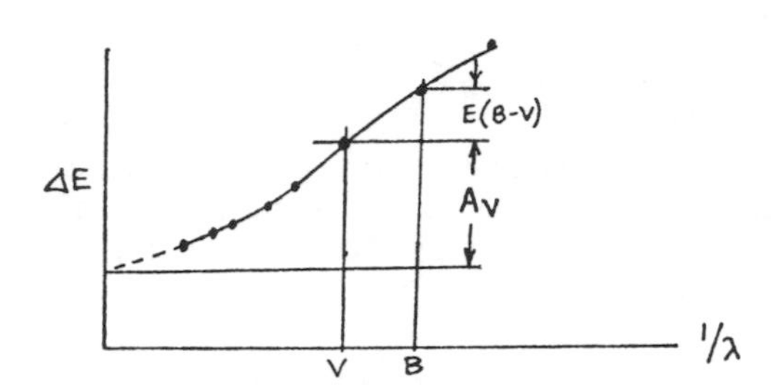

In the visual wavelength range, the extinction follows roughly a $1/\lambda$ wavelength dependence.

(b) Assume that all stars of the same spectral class have the same intrinsic luminosity:

$$m_\lambda - M_\lambda = 5\ (\log r - 1) + A_\lambda$$

where M_λ is the absolute magnitude, m_λ the observed magnitude, r the distance in parsecs, and A_λ the total extinction.

Both of these methods are subject to large uncertainties due to the assumptions involved and because the involved quantities (e.g. M_λ, r) are not well-known. Other methods also have been used.

The best general value at present is

$$A_V = 3E(B-V)$$

but $A_V/E(B-V)$ ratios as high as 6 have been indicated in some regions of the sky.

The Orion nebula, in particular, seems peculiar in that the E(B-V) value is less than would be indicated by the rest of the extinction curve and the estimated total extinction.

Rocket and satellite observations have enabled us to extend the extinction curves into the far-ultraviolet and have provided much new information which may help us to understand the properties and composition of dust grains.

Early in the history of rocket ultraviolet astronomy, there were indications of a peak in the extinction curve near 2200 Å, which seemed to be characteristic of graphite.

Therefore, graphite quickly became a popular basis for models of interstellar grains.

The OAO-2 Wisconsin experiment observations confirmed the presence of this peak, at nearly the same wavelength for all star pairs (one reddened/one unreddened) observed, but with different amplitudes relative to A_V.

Shortward of the "graphite peak" there were even larger variations among the star pairs observed; although, in general, the behavior was that the extinction decreased to a minimum and then continued upward below about 1500 Å. As indicated by ground-based observations, ϕ Ori was especially peculiar.

Fig. 12 shows a summary of OAO-2 extinction measurements. Recent measurements by OAO-3 have extended the extinction curves down to 1000 Å and have shown that the

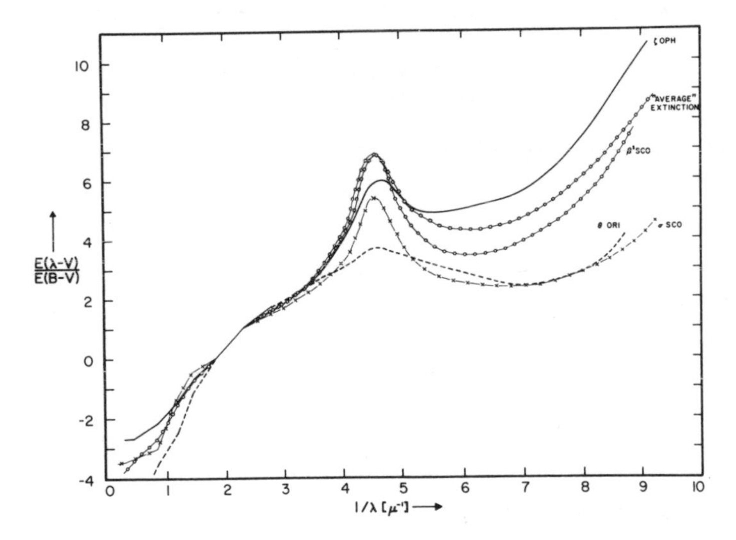

Fig. 12: Interstellar extinction curves for various stars as observed by OAO-2, normalized to E(B-V)=1 (Bless and Savage, 1972).

extinction continues to increase toward shorter wavelengths in this range. A predicted feature due to graphite at 1030 Å was not definitely observed. For ζ Oph and ζ Per, the increase of one, in inverse micron units corresponds to an extinction increase of 3 magnitudes for an E(B-V)=1.

Most of the observed properties of the interstellar

dust are satisfied by a recent grain model (Gilra, 1971) based on mixtures of graphite, silicon carbide, and meteoric silicate particles.

2. Reflection nebulae and galactic background

Reflection nebulae are the result of scattering of starlight by dust grains. The general galactic background light is also believed to be the result of scattering in the general interstellar dust.

The best example of a reflection nebulosity is that associated with the Pleiades. Another example is Barnard's loop, a ring of reflection nebulosity surrounding the Orion complex of early-type stars.

Study of the galactic background light is difficult from the ground because of stray light due to airglow, atmospheric scattering, unresolved faint stars, and zodiacal light. The latter two are still present in space observations, but various methods can be used to remove or correct for these conbributions.

A model of the interstellar dust grains proposed by Witt (1972) to explain the OAO observations consists of the following:

(1) A few micron-sized, ice-coated particles which dominate scattering in vis. and near UV.

(2) Numerous sub-micron, graphite-like grains which absorb very strongly near 2200 Å.

(3) Large numbers of sub-micron, silicate-like grains which dominate scattering in the far UV.

Photometry of the sky background and reflection nebulosities have been carried out in the 1500-4000 Å wavelength range by OAO-2. Witt and Lillie (1972) found that, in the general interstellar medium, the albedo of the grains was fairly constant at a value of about 0.5 in the 3000-4000 Å range, then dropped off rapidly to a minimum of about 0.1 near 2200 Å. The albedo then rose toward shorter wavelengths, reaching a value near unity at 1500 Å (see Fig. 13). Ground-based observations of the general galactic background have indicated an albedo of about 0.65 in the B and V bands. However, comparisons of ground-based observations with OAO-2 and OSO-6 photometry (Roach and Lillie, 1972) indicate that the ground-based measurements

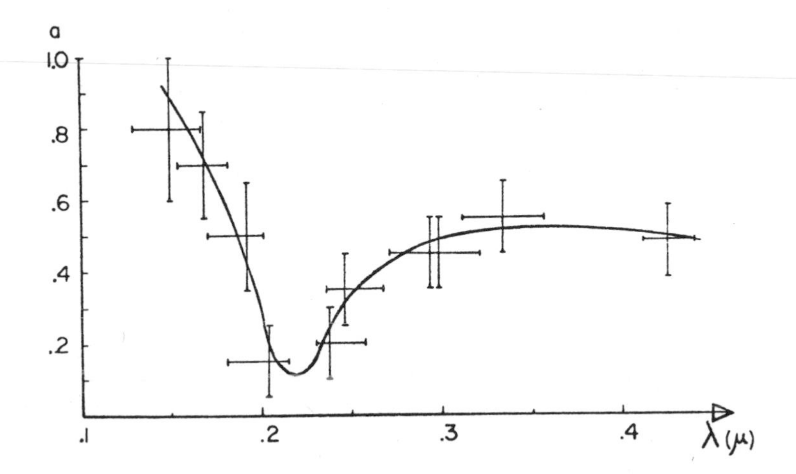

Fig. 13: Albedo curve for dust grains in the general interstellar medium, as obtained from OAO-2 observations (Witt and Lillie, 1972).

of the sky background are too high, perhaps due to previously uncorrected airglow emissions.

Ground-based observations of the Pleiades nebulosity have indicated that the reflected light is somewhat bluer than that of the exciting stars. Lillie and Witt (1972) have obtained considerably more detail in their OAO-2 observations (Fig. 14). The minimum in the albedo curve

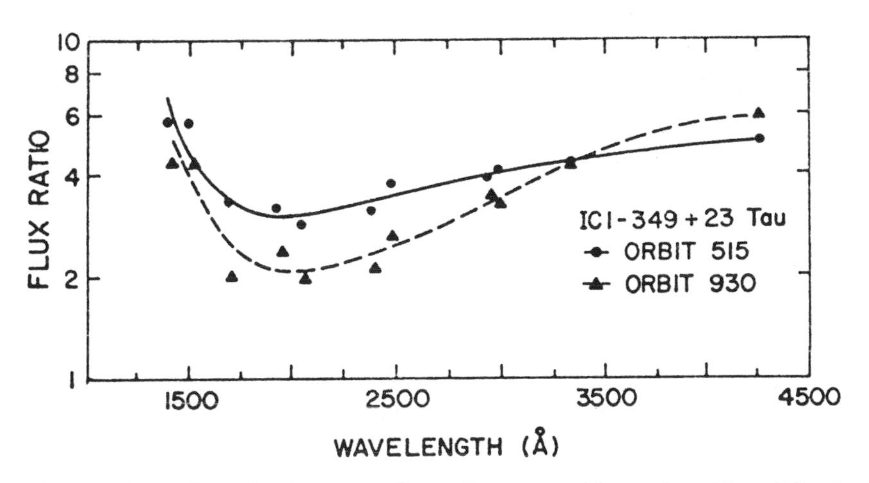

Fig. 14: Albedo curve for dust grains in the Pleiades reflection nebulosity according to OAO-2 (Lillie and Witt, 1972).

is much less pronounced, or even absent, compared to the general interstellar reflection curve. This is interpreted as a local depletion of the small graphite particles believed to be responsible for the 2200 Å absorption.

Barnard's loop was observed in the 2200-5000 Å wavelength range by O'Dell, York, and Henize (1967). It was found to be uncexpectedly bright in the UV compared with the visible, indicating increasing grain albedo in the near UV (shortward of the graphite absorption). Lillie and Witt find their measurement at 1500 Å to be consistent with O'Dell et al. in Barnard's loop, assuming the same reflectivity curve as for the Pleiades. However, an NRL experiment in 1968, using an electronographic Schmidt camera (Henry and Carruthers, 1970), sensitive in the 1230-2000 Å wavelength range, failed to detect Barnard's loop. It is possible that the camera was not sensitive enough with the exposure time used, and we plan a repeat observation.

REFERENCES

Bates, D., Spitzer, L., Ap.J. 113, 441, 1951.

Bless, R.C., Savage, B.D., Ap.J. 171, 293, 1971.

Boer, K.S. de, Hoekstra, R., Hucht, K.A. van der, Kamperman, T.M., Lamers, H.J., Pottasch, S.R., Astr. and Ap. 21, 447, 1972.

Bortolot, V.J.,Jr., Clauser, J.F., Thaddeus, P., Phys. Rev.Lett. 22, 307, 1969.

Bortolot, V.J., Jr., Thaddeus, P., Ap.J.Lett. 155,L17, 1969.

Carlson, R.W., private communication, 1972.

Carruthers, G.R., Ap.J. 151, 269, 1968.

Carruthers, G.R., Ap.J.Lett. 156, L97, 1969.

Carruthers, G.R., Ap.J. 166, 348, 1971.

Carruthers, G.R., Bromberg,K.L., Bull.Am.Astr.Soc. 4, 269, 1972.

Carruthers, G.R., Page, T.L., in Apollo 16 Preliminary Science Report, NASA Sp-315, p.13, 1972.

Dufay, J., Galactic Nebulae and Interstellar Matter, Dover, New York, 1968.

Field, G.B., Somerville, W.B., Dressler, K., in Ann.Rev. Astron.Astrophys. (L. Goldberg, ed.), Annual Reviews Inc., Palo Alto, Calif., 1966.

Frisch, P., Ap.J. 173, 301, 1972.

Gilra, D.P., Nature 229, 237, 1971.

Henry, R.C., Carruthers, G.R., Science 170, 527, 1970.

Herbig, G.H., J.Q.S. and R.T. 3, 529, 1963.

Herbig, G.H., Zs.f.Ap. 68, 243, 1968.

Herbig, G.H., review paper presented at June meeting of A.A.S., Boulder, Col., 1970.

Jenkins, E.B., in Ultraviolet Stellar Spectra and Related Ground Based Observations (IAU Symposium No. 36), D. Reidel, Dordrecht, p. 281, 1970.

Jenkins, E.B., Ap.J. 181, 761, 1973.

Johnson, H.L., in Nebulae and Interstellar Matter (B.M. Middlehurst and L.H. Aller, eds.) Univ. of Chicago Press, p. 191, 1968.

Kaplan, S.A., Pikelner, S.B., The Interstellar Medium, Harvard Univ. Press, Cambridge, Mass., 1970.

Lillie, C.F., Witt, A.N., presented at IAU Symposium No. 52, Interstellar Dust and Related Topics, Albany, New York, 1972.

Morton, D.C., Ap.J. 147, 1017, 1967.

Morton, D.C., Jenkins, E.B., Brooks, N.H., Ap.J. 155, 875, 1969.

Morton, D.C., Jenkins, E.B., Matilsky, T.A., York, D.G., Ap.J. 177, 219, 1972a.

Morton, D.C., Jenkins, E.B., Macy, W.W., Ap.J. 177, 235, 1972b.

Munch, G., in Nebulae and Interstellar Matter (B.M. Middlehurst and L.H. Aller, eds.) Univ.of Chicago Press, p. 365, 1968.

O'Dell, C.R., York, D.G., Henize, K.G., Ap.J. 150, 835, 1967.

Osterbrock, D.E., Planet.Sp.Sci. 11, 621, 1963.

Pottasch, S.R., in Vistas of Astronomy (A. Beer, ed.), Pergamon Press, New York, Vol. 6, p. 149, 1965.

Roach, F.E., Lillie, C.F., in The Scientific Results from the Orbiting Astronomical Observatory (OAO-2), NASA Sp-310, p. 71, 1972.

Savage, B.D., Jenkins, E.B., Ap.J. 172, 491, 1972.

Smith, A.M., Ap.J. 160, 595, 1970.

Smith, A.M., Ap.J. 172, 129, 1972a.

Smith, A.M., Ap.J. 176, 405, 1972b.

Smith, A.M., Stecher, T.P., Ap.J.Lett. 164, L43, 1971.

Solomon, P.M., Klemperer, W., Ap.J. 178, 389, 1972.

Spitzer, L., Diffuse Matter in Space, John Wiley & Sons, New York, 1968.

Stecher, T.P., Williams, D.A., Ap.J.Lett. 149, L29, 1967.

Stone, M.E., Morton, D.C., Ap.J. 149, 29, 1967.

Witt, A.N., presented at IAU Symposium No. 52, Interstellar Dust and Related Topics, Albany, New York, 1972.

Witt, A.N., Lillie, C.F., in The Scientific Results from the Orbiting Astronomical Observatory (OAO-2), NASA SP-310, p. 199, 1972.

Spectrophotometric Results from the Copernicus Satellite (Ap.J.Lett., Vol. 181, 1973)

Rogerson, J.B., Spitzer, L., Drake, J.F., Dressler, K., Jenkins, E.B., Morton, D.C., York, D.G., I. Instrumentation and Performance, L97.

Morton, D.C., Drake, J.F., Jenkins, E.B., Rogerson, J.B., Spitzer, L., York, D.G., II. Composition of Interstellar Clouds, L103.

Rogerson, J.B., York, D.G., Drake, J.F., Jenkins, E.B., Morton, D.C., Spitzer, L., III. Ionization and Composition of the Intercloud Medium, L110.

Spitzer, L., Drake, J.F., Jenkins, E.B., Morton, D.C., Rogerson, J.B., York, D.G., IV. Molecular Hydrogen in Interstellar Space, L116.

Jenkins, E.B., Drake, J.F., Morton, D.C., Rogerson, J.B., Spitzer, L., York, D.G., V. Abundances of Molecules in Interstellar Clouds, L122.

In press: (Ap.J.Lett.)

York, D.G., Drake, J.F., Jenkins, E.B., Morton, D.C., Rogerson, J.B., Spitzer, L., VI. Extinction by Grains at Wavelengths between 1200 and 1000 A.

THE INTERSTELLAR MEDIUM IN X- AND GAMMA RAY ASTRONOMY

C. Dilworth

Università di Milano
Milan, Italy

I. THE PRESENT STATUS OF X- AND GAMMA RAY ASTRONOMY

X- and gamma ray astronomy is a statistics-limited type of astronomy. The number of sources known is small, less than 200, and although the amount of energy received per unit area in unit time is large, the number of photons involved is, in general, small, typically of the order of $0.1/cm^2$ sec for X-rays down to $10^{-4}/cm^2$sec for gamma rays. In the range of wavelength covered, from 10^{-2} Å to 100 Å, five main intervals, investigated by different techniques and subject to different problems, can be distinguished. The characteristics of these five regions are summarized in Table I, in which are noted the experimental techniques most commonly employed, values of the counting-rate observed, or expected, in a typical instrument, for a source such as Cygnus X-2 and the background above which such a source has to be seen.

The physical mechanisms leading to the absorption, and hence detection, of X- and gamma rays, are the photoelectric effect, Compton scattering and pair production. The cross-section for these three processes are very different functions of the atomic number, Z, of the absorbing material and of the energy, $h\nu$, of the photon (Table II).

The energy region in which the cross-section for detection is highest (the soft X-ray region), is also that in which interstellar absorption is strongest. In

	SOFT X-RAYS $0.1 \lesssim E \lesssim 1$ KeV $120 \gtrsim \lambda \gtrsim 12$ Å	MEDIUM X-RAYS $1 \lesssim E \lesssim 10$KeV $12 \gtrsim \lambda \gtrsim 1$ Å	HARD X-RAYS $10 \lesssim E \lesssim 100$KeV $1 \gtrsim \lambda \gtrsim 0.1$ Å	LOW ENERGY γ-RAYS 100KeV $\lesssim E \lesssim$ 10MeV $0.1 \gtrsim \lambda \gtrsim 10^{-2}$ Å	HIGH ENERGY γ-RAYS $E \gtrsim 10$ MeV $\lambda \lesssim 10^{-2}$ Å
Detector	Gas-flow proportional Counter (~100 μgm/cm2 window) or electron multipliers	Proportional Counters (10^{-2} gm/cm^2 window)	Scintillation Counters	Scintillation Counters	Spark Chambers
Directional Element	Mechanical collimators or Grazing incidence optics	Collimators	Active anti-coincidence collimators	Active anti-coincidence collimators	Visual image of γ-ray pair
Vehicle	Rocket or satellite	Rocket or satellite	Balloon, rocket or satellite	Balloon or satellite	Balloon or satellite
Typical source Count rate	3 c/cm^{2}sec	0.5c/cm^{2}sec	0.2 c/cm^{2}sec	10^{-2}c/cm^{2}sec	$< 10^{-4}$c/cm^{2}sec
Typical Background count rate	1 c/cm^{2}sec	10^{-2}c/cm^{2}sec	0.1 c/cm^{2}sec	2 10^{-1}c/cm^{2}sec	2 10^{-4}c/cm^{2}sec
Typical Angular Resolution	~ 1°	~ 1 arc min	~ 6°	~ 10°	~ 2°

Table I: Characteristics for the investigation of photons in five different energy regions.

Mechanism	Cross-section cm^2	Mass attenuation coefficient (cm^2/g) at	10 KeV	1 MeV	100 MeV
Photoelectric Effect	$\sim NZ^4/(h\nu)^3$	In Air	4	$< 10^{-3}$	$< 10^{-3}$
		In Pb	80	$2\ 10^{-1}$	$< 10^{-3}$
Compton Scattering (total)	$\frac{8\pi}{3} NZ\, r_e^2 \left[\frac{1+\alpha}{\alpha^2} \left\{ \frac{2(1+\alpha)}{1+2\alpha} - \frac{1}{\alpha} \ln(1+2\alpha) \right\} + \frac{1}{2\alpha} \ln(1+2\alpha) - \frac{1+3\alpha}{(1+2\alpha)^2} \right]$	In Air	$2\ 10^{-1}$	$6.5\ 10^{-1}$	$2\ 10^{-3}$
		In Pb	$1.5\ 10^{-1}$	$5\ 10^{-1}$	$1.5\ 10^{-3}$
Pair Production	$N\sigma_0 Z^2 (\frac{28}{9} \ln 2\alpha - \frac{218}{27})$ for $I < \alpha \ll 137\ Z^{-1/3}$	In Air	-	-	2.10^{-2}
	$N\sigma_0 Z^2 \left\{ \frac{28}{9} \ln(187\ Z^{-1/3}) - \frac{2}{27} \right\}$ for $\alpha \gg 137\ Z^{-1/3}$	In Pb	-	-	1.10^{-1}

$\alpha = h\nu/m_e c^2$; r_e = classical radius of electron; N=n° of atoms/c.c.; Z= atomic n°;

$\sigma_0 = \frac{1}{137} r_e^2$.

Table II: Attenuation of photons in matter.

the medium X-ray region the detection efficiency is still high, but interstellar absorption is limited, and it is in this range that the most complete coverage is available at present. Most, but not all, of the sources seen so far in soft X-rays are visible also in the medium X-ray range. No systematic scan in the hard X-ray or gamma ray regions has yet given firm evidence for additional sources.

From the UHURU catalogue (Giacconi et al., 1972) which gives the most complete survey of medium X-ray sources to date, a rough separation into galactic and extragalactic sources can be made on the basis of their distribution in galactic latitude. Sources with $|b^{o}| < 20^{o}$ are on the average more intense, and their log N - log S distribution approaches more closely that expected for a disk, while those at high latitude, $|b| > 20^{o}$ are on the average weaker, with a log N - log S distribution compatible with the $S^{-1.5}$ law expected from spherical symmetry (Matilsky et al., 1972) (Fig. 1). A strong clustering of the most intense sources in the longitudinal sector, $330^{o} < l < 30^{o}$, around the direction to the galactic center has led to the postulation of a separate class of high luminosity sources concentrated in this region (Salpeter, 1972; Gursky, 1972; Dilworth et al., 1973).

Studies of the time variation of the intensity of the sources lead to a division in four main classes

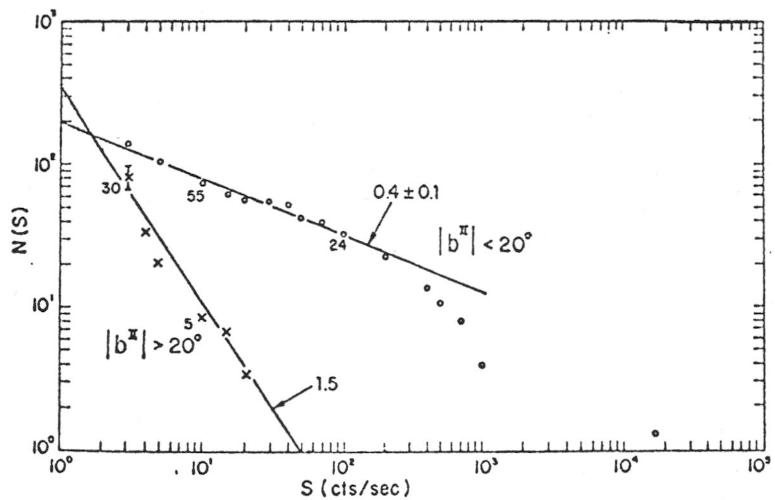

Fig. 1: X-ray source number distribution corrected for sky coverage (Gursky, 1972).

(Tananbaum, 1973):

(1) Fast aperiodic variables
(2) Novae
(3) Periodic variables (pulsars; binary systems)
(4) Steady non varying sources.

In classes (1) and (3) one finds compact objects near to the end of their evolutionary path. In class (4) are to be found supernovae remnants, some of which (e.g. Cygnus loop) are not in the UHURU catalogue, being of low temperature and hence visible only in the soft X-ray frequency range.

Spectral measurements on X-ray sources are far from being complete or detailed. Only on the very brightest sources (e.g. SCO-X1 and Crab) are measurements available over the whole range from soft to hard X-rays. The energy resolution obtainable with proportional counters is poor (at best $\frac{\Delta E}{E} \sim 10\%$) and the sensitivity response of the counter so far from uniform that spectral curves cannot be derived directly. A form of the spectrum must be assumed, the relative expected count-rate deduced and compared to observations. On the variable sources in which both intensity and spectrum can vary, simultaneous measurements are necessary (but not always possible) in all frequency bands.

In general, one of two types of spectra is assumed either a thermal, $I(E) = \frac{I}{E} e^{-E/kT}$ +), or a power law, $I(E) = KE^{-\alpha}$, spectrum. In some cases (e.g. SCO-X1, Cyg 1) a thermal spectrum in the soft and medium range is found to have a power law tail in the hard X-ray region.

In all the frequency ranges, in addition to the resolved sources, a "background" radiation is observed. To what extent this background radiation can be attributed to the integrated emission of unresolved sources and to what extent it may be due to a truly diffuse emission is an as yet unresolved problem. The problem is at a very different level in the medium X-ray region, where over

---

+) A more correct form (Chodel et al., 1968) is

$$I(E) = \frac{K}{E} e^{-E/2kT} G(E/2kT)$$

$$\text{where } G(E/2kT) \simeq \begin{cases} \ln(4kT/E) - 0.577 & \text{for } E/2kT \ll 1 \\ (\pi/2)^{1/2} e^{-E/2kT} (E/2kT)^{-1/2} & \text{for } E/2kT \gg 1 \end{cases}$$

100 sources have been resolved and the galactic background (~ 10% of the isotropic background) is barely visible, and in the high energy gamma ray region in which practically all the firm data consists of measurement of background. We will discuss this problem separately therefore in the chapter on diffuse emission.

II. ABSORPTION AND SCATTERING

A glance at the absorption curves of photons in diverse elements tells us that for energies above ~10 keV the attenuation coefficient is well below 100 cm^2/gm in all but the heaviest elements, thus absorption in the interstellar medium will be of importance only in the soft and medium X-ray range. The mechanism operative in this energy region is the photoelectric effect, whose cross-section is strongly dependent both on energy and on Z, the atomic number of the absorbing material (see Table II), with sharp changes at the K and L thresholds.

The effect of photoelectric absorption is seen in the energy spectrum as a fast fall in intensity at low energies. If I(E) is the unabsorbed spectrum of the source, the observed spectrum will be:

$$\bar{J}(E) = I(E)\, e^{-\sigma N_H}$$

where N_H is the columnar density of gas expressed in number of hydrogen atoms, and σ the absorption cross-section relative to the atomic constitution of the gas and to the energy considered. A "cut-off"energy, E_a, can be defined as that energy at which the observed intensity falls to 1/e of that defined by I(E): i.e. $\sigma N_H = 1$ and $\bar{J}(E_a) = 1/e\ I(E_a)$. Knowing the form of the spectrum I(E) at energies $E > E_a$, one can deduce, from the ratio of the observed intensity to that expected by extrapolation, the absorption factor $e^{-\sigma N_H}$.

In practice of course the atomic constitution of the gas is not well known nor is the energy spectrum of the source measured to the desired accuracy. The fitting of the measured count-rate spectrum, with generally low statistics, by a function which folds in the response curve of the detector to a spectrum deformed by absorption, leads to large uncertainties in the parameters derived. Examples of this problem are given in Figs. 2 and 3 from Bleeker et al. (1972) which show the type of fit which is obtained and the range of values of N_H and kT leading to a reasonable value of χ^2. Furthermore it

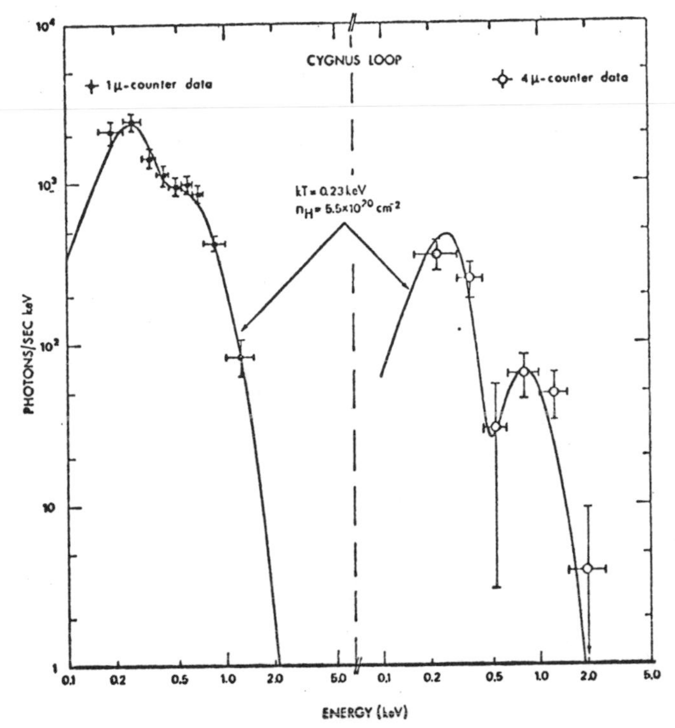

Fig. 2: (Bleeker et al., 1972) Energy spectrum for the Cygnus loop obtained with 1μ window (solid circles) and 4μ window (open circles) counters. The solid lines represent the best fit to a thermal bremsstrahlung source with a temperature of 2.7x10^6 K with interstellar absorption by cool material of 5.5x10^{20} H atoms/cm^2.

is not always evident whether absorption is taking place in the interstellar medium or in the source itself.

Considerable work has been carried out on the problem of interstellar absorption for, in the absence of optical identification, it is at present the only distance indicator one has for the majority of X-ray sources. Brown and Gould (1970) have computed the effective cross-section for the interstellar gas using the abundances indicated in Table III. As can be seen from their curves (Fig. 4)

Element	H	He	C	N	O	Ne	Mg	Si	S	Ar
$\log_{10}$ N	12.00	10.92	8.60	8.05	8.95	8.00	7.40	7.50	7.35	6.88

Table III: Abundances,N, in the interstellar medium assumed for calculation of interstellar absorption (Brown and Gould, 1972).

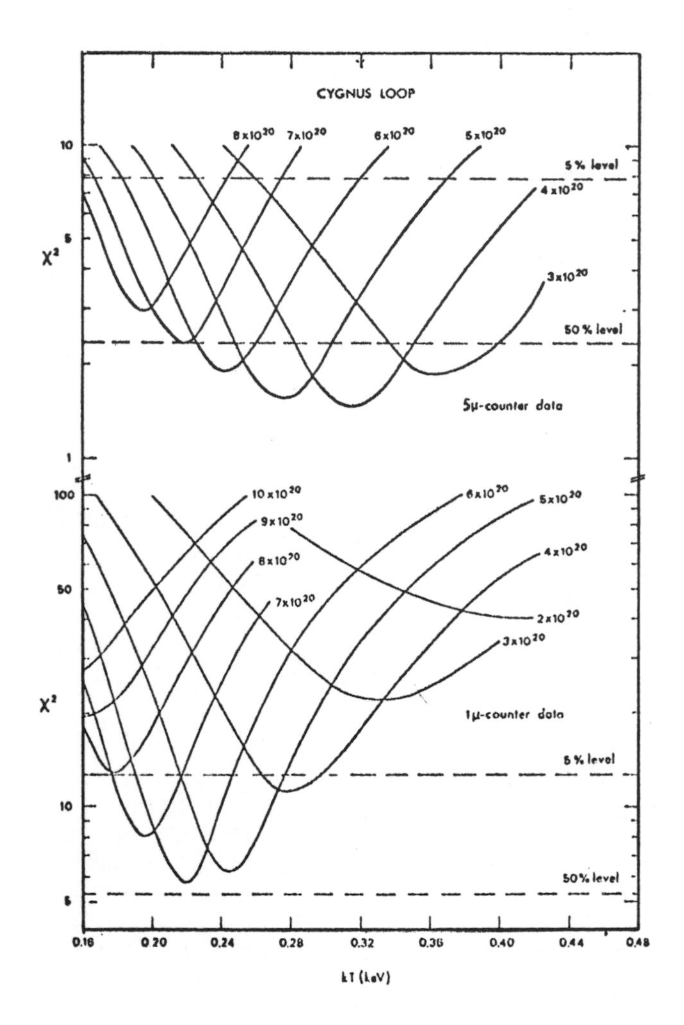

Fig. 3: (Bleeker et al., 1972) χ^2-values versus plasma temperature for various values of the parameter N_H. The upper set of curves applies to the 4μ window counter data, the lower set to the 1μ window counter data.

elements of higher atomic number become effective as the energy of the X-rays increases beyond the relative K-level. Since the energy of cut-off, E_a, of the spectrum is higher for the more distant sources, the effective cross-section measured for distant sources is mainly due to the heavier elements (Fig. 5). An approximate expression smoothing out the K-edge irregularities is (Hayakawa, 1973):

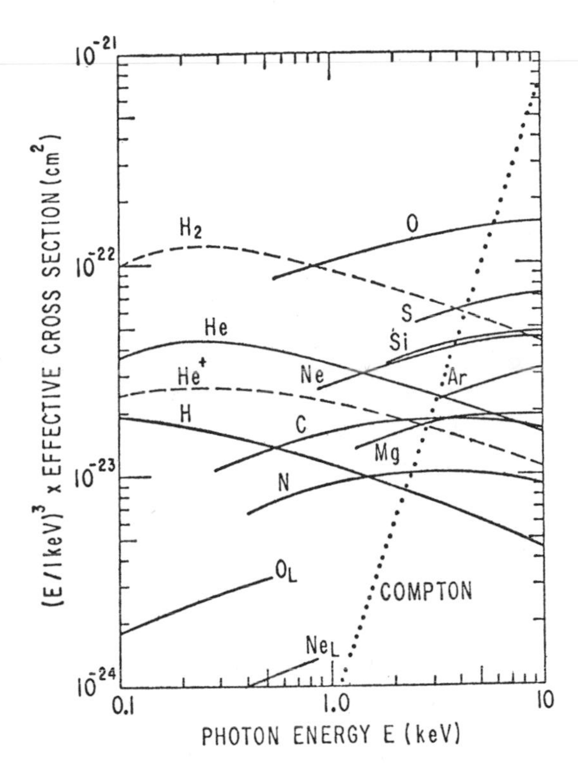

Fig. 4: (Brown and Gould, 1970) Variation with photon energy of photoelectric absorption cross-section of constituents of interstellar matter.

$$\sigma(E) = \begin{cases} 6.0\times10^{-20}(0.1\text{kev}/E)^{3}\text{cm}^{2} & \text{for } 0.1<E<0.53 \text{ keV} \\ 2.0\times10^{-22}(1 \text{ keV}/E)^{2.5}\text{cm}^{2} & \text{for } 0.53<E<10 \text{ keV} \end{cases}$$

Using previous (Bell and Kingston, 1967) effective cross-sections higher than these, Ilovaisky (1971) has compared X-ray absorption with 21-cm radio emission along the line of sight to the source and found some tendency for an overestimate of N_H from X-ray absorption. Reina and Tarenghi (1973), however, comparing X-ray absorption and optical extinction $A_V = A$ (5500 Å) suggest for 4 identified objects the relation

$$A_V = 5.4 \times 10^{-22} N_H \quad \text{(magnitudes)}$$

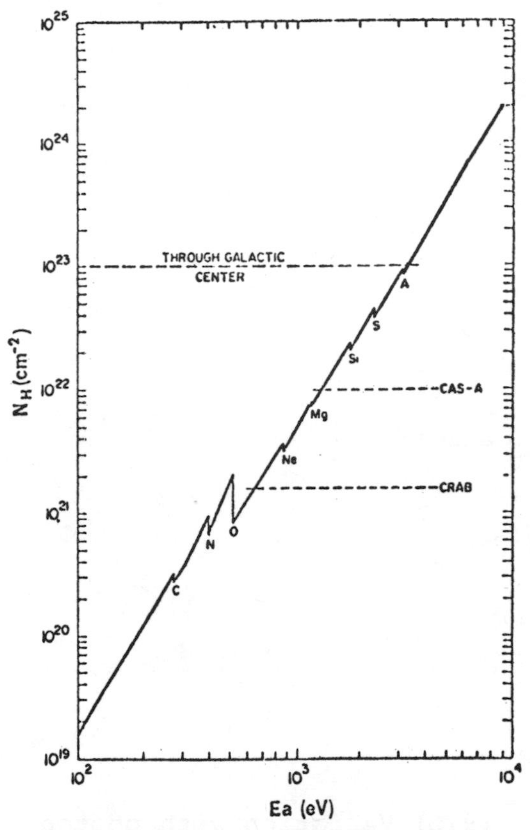

Fig. 5: (Gursky, 1972) Photoelectric absorption in the interstellar medium. The vertical axis gives the column density in units of hydrogen atoms/cm^2 at which the absorption is 1/e at the appropriate photon energy E_a.

between X-ray defined N_H and optical extinction A_V and note that this relation is in good agreement with that derived by Lilley (1955) from radio absorption.

Seward et al. (1972) have measured the photoelectric cut-off in 30 sources, and on the basis of known (through optical identification) or deduced (position with respect to the galactic center) distance, established a "distance indicator" in the form of a curve which relates distance in kpc and N_H calculated with the Brown and Gould cross-sections (Fig. 6). They suggest that the deviation of their curve from simple proportionality could be due to an increasing concentration of heavy elements towards the galactic center.

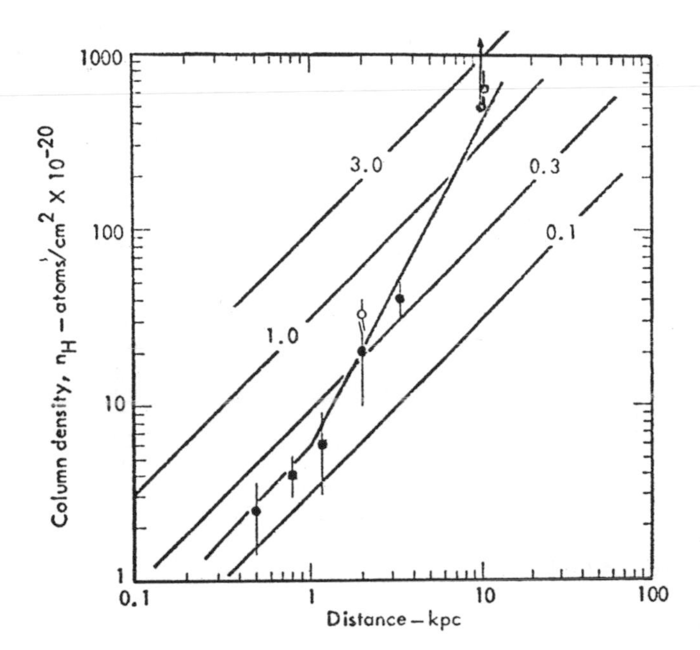

Fig. 6: (Seward et al., 1972) Absorption versus distance calibration curve. Data points are, in order of increasing distance, Vela X, Cygnus loop, Puppis A, Crab nebula, Cas A and the galactic center source. Solid circles - Seward et al. (1972); open circles - Henry et al. (1972) for the Crab and Kellogg et al. (1971) for the galactic center source. The arrows show lower limits. The sloping lines are lines of constant density, and the number indicate atoms per cm^3.

This brings one to a hope for the future. As Brown (1972) has pointed out, the strong dependence on Z and $h\nu$ of the photoelectric cross-section and the presence of the K absorption edges should introduce a recognizable pattern into an otherwise smooth energy spectrum, thus offering a means of measuring the relative composition of the interstellar medium (Table IV). Hayakawa and Tanaka (1973) have in fact suggested that irregularities in the Cen X-3 spectrum can be attributed to the K-edges of oxygen, silicon and iron, in this case present in the highly ionized circumstellar envelope of the source.

How dependable an instrument for the investigation of the interstellar medium X-ray absorption measurements can be made, will depend both on the increase in precision which can be attained and on the degree to which

Atom	Abundance ($N_H = 1$)	*K*-Edge (KeV)	σ_0 (10^{-19} cm^2)	$\log N_H$ ($\tau = 1$)	$\Delta\phi$ (%)			
					log N_H (cm^{-2}) 22.5	22.0	21.5	21.0
C	3×10^{-4}	0.284	10.4	20.17	100	95.6	62.7	26.8
N	1×10^{-4}	0.400	16.8	20.76	99.5	81.4	41.2	15.4
O	7×10^{-4}	0.532	6.30	21.12	100	98.8	75.2	35.7
F	2.5×10^{-7}	0.686	4.16		0.033	0.010	0.003	0.001
Ne	1×10^{-4}	0.867	3.20	21.51	63.6	27.4	9.62	3.14
Na	1.7×10^{-6}	1.072	2.55		1.39	0.44	0.140	0.040
Mg	3×10^{-5}	1.303	1.88	21.96	16.3	5.48	1.77	0.560
Al	2×10^{-6}	1.560	1.57		0.98	0.31	0.096	0.032
Si	3×10^{-5}	1.840	1.38	22.38	12.3	4.05	1.30	0.413
P	3.5×10^{-7}	2.143	1.16		0.13	0.041	0.013	0.004
S	1.7×10^{-5}	2.470	0.984	22.69	5.14	1.66	0.527	0.167
Cl	2.5×10^{-7}	2.820	0.862		0.065	0.020	0.006	0.003
A	4.2×10^{-6}	3.203	0.741	22.91	0.98	0.310	0.098	0.031
K	7.5×10^{-8}	3.608	0.651		0.015	0.005	0.002	0.001
Ca	1.7×10^{-6}	4.038	0.571		0.306	0.100	0.030	0.010

Table IV: K-edge spectral discontinuities (Brown, 1972).

knowledge of the nature of the sources will allow one to separate interstellar from self-absorption.

Another hope for the future is that of the observation of the effect of Thompson scattering at low energies by interstellar grains which is expected to give rise to "halos" around the sources. That such halos have not yet been observed is not surprising, as the present instrumentation based on collimated proportional counters does not have the necessary angular resolution nor sensitivity. Typical halos can be expected to be of the order of arc minutes in diameter, and to contain 20% of the intensity of the source. In the future, with grazing incidence telescopes (Giacconi and Rossi, 1960) or lunar occultation observations (Collet et al., 1970) these phenomena should be observable.

Some indication of the appropriateness of given grain models could possibly be obtained from the observed profile of the halo (Martin, 1970). In addition, as Trümper and Schönfelder (1973) suggest, observations of the "damping" in the halo of time variations in the source can determine the geometrical distance of the source and hence one could in principle obtain direct information on the grain density.

III. DIFFUSE EMISSION BY THE INTERSTELLAR MEDIUM

1. Emission mechanisms. The emission of X- or gamma radiation by the interstellar medium can occur through thermal bremsstrahlung in a hot interstellar plasma, bremsstrahlung, synchrotron emission or inverse Compton effect of cosmic ray electrons, or the decay of π^o-mesons produced in the interaction of cosmic ray nucleons with interstellar matter.

The possibility of gravitational heating of intergalactic gas falling into the galaxy to a temperature of $\sim 10^6$ ^oK required for the emission of soft X-rays, has been discussed by Hunt and Sciama (1972).

The role of cosmic ray electrons in the production of non-thermal radio background by synchrotron emission has been discussed in previous lectures by P.G. Mezger.

The relative importance of bremsstrahlung, inverse Compton and synchrotron emission in the production of X- and gamma radiation is not simple to estimate (Ginzburg, 1964) since the energy of the electrons required for the emission of a photon of a given energy differs by several orders of magnitude in the three mechanisms (Fig. 7) covering a range of energies broader than that for which the intensity of cosmic ray electrons is known to any degree of confidence. The intensity of emission per electron of photons of a given energy, E_{ph}, can be readily calculated (Table V). To deduce the total galactic emission one needs to make assumptions as to the flux of electrons of corresponding energy, E, on the basis of extrapolation and demodulation of the cosmic ray electron spectrum observed near the earth.

The estimate of the contribution of π^o-meson decay to the flux of gamma rays of $\sim$ 100 MeV is subject to less uncertainties. The cross-sections for production of π^o-mesons, their energy spectrum and life-time for decay into 2 gamma rays, are known from measurements at accelerators, and the flux of cosmic ray nucleons is reasonably well known. With this data, calculations by different authors, Cavallo and Gould (1971) and Stecker (1970) lead to estimates of the gamma ray flux which differ by less than a factor two.

2. Experimental data. As pointed out in section I, it is extremely difficult to discuss experimental data on the diffuse emission until we have sufficient information to estimate the contribution of unresolved

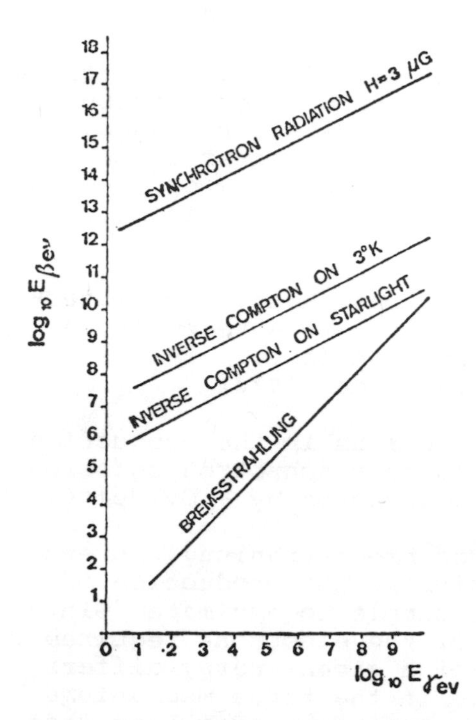

Fig. 7: Relation between energy of electron, E_β , and average energy, E_γ , of photon radiated by synchrotron emission in a 3 microgauss magnetic field, by inverse Compton effect on the 3^oK black body universal radiation and on starlight and by bremsstrahlung.

sources. Since the relative total intensity of resolved sources is very different in the various energy bands, we will discuss the problem separately for each.

A. The soft X-ray region

Photoelectric absorption in the interstellar gas is so strong in this region that all but the nearest of the medium X-ray sources are obscured. On the other hand, a few very low temperature sources, mostly identified with supernovae remnants appear here, and not in the medium X-ray band. A complete satellite survey is not yet available, restricted rocket scans show a broad irregular X-ray emission with a few isolated peaks corresponding to the resolved sources.

The variation with latitude of the intensity of this

Energy loss per unit time by an electron of energy E	Average Energy of photon radiated by electron of energy E	Energy of electron required to produce photons of average energy E_γ	Intensity of photons of Energy E_γ radiated per electron
dE/dt	E (KeV)	E (eV)	I (KeV/sec)
I Synchrotron Emission			
$\frac{32\pi}{9} c\, r_e^2 (\frac{H}{8\pi})^2 (\frac{E}{mc^2})^2$	$0.07\ \frac{eh}{mc} H_{\perp} (\frac{E}{mc^2})^2$	$2.2\ 10^{19} (\frac{E_\gamma}{H_{\mu G}})^{1/2}$	$0.2\ H_{\mu G}\ E_\gamma$
II Inverse Compton Emission			
$\frac{32\pi}{9} c\, w_{ph} r_e^2 (\frac{E}{mc^2})^2$	$\frac{4}{3} \bar{\varepsilon} (\frac{E}{mc^2})^2$	$3.75\ 10^5 (\frac{E_\gamma}{\bar{\varepsilon}})^{1/2}$	$2.6\ 10^{-14}\ \frac{w_{ph}}{\bar{\varepsilon}}\ \frac{3}{4}\ E_\gamma$
III Bremsstrahlung			
$8.0\ 10^{-16}\ n \cdot E$	0.5 E	$2\ E_\gamma$	$2\ 10^{-16}\ E_\gamma$

r_e = classical radius

H = average magnetic field (oersted)

mc^2 = electron rest mass

w_{ph} = energy density of photons

n = density of interstellar hydrogen

$H_{\mu G}$ = magnetic field in micro oersted.

$\bar{\varepsilon}$ = average energy of photons

Table V: Radiation of photons by relativistic electrons.

background radiation has given rise to much discussion. Due to the difficulty of separating the true X-ray emission from spurious background effects (due to UV and particle contamination), discrepancies occurred between different observations. Improvement in experimental techniques has led more recently to some agreement on the experimental picture, which, however, remains complex in its nature.

Originally attempts were made to interpret the soft X-ray background as due to an isotropic extragalactic emission (similar to that observed in medium X-rays), absorbed by the interstellar medium, thus giving higher intensity towards the galactic poles than toward the galactic plane. Such an effect is observed in the direction of the anticenter ($90^o < l < 270^o$). However, the observation of patches of strong intensity at low latitudes in the forward longitude zone, and the failure to observe a decrease in the emission when scanning across the small Magellanic cloud (McCammon et al., 1971) makes it difficult to accept a purely extragalactic origin.

Recent analyses (Friedman et al., 1973; Hayakawa, 1973) have tended therefore to suggest the presence of both galactic and extragalactic emission or alternatively of a galactic emission in a disk of height much greater than that of the interstellar gas.

The high level of the granularity, of the background (Gorenstein and Tucker, 1972) at this stage seems to indicate that the greater part of it should be due to as yet unresolved sources. If considered as diffuse emission one would need a mechanism producing temperatures as high as 10^6 ^oK.

B. The medium X-ray region

The dominant background in this region is isotropic, and hence presumably extragalactic in origin. Recent measurements (Cooke and Griffiths, 1969; Bleach et al., 1972; Gursky, 1972) have shown the existence of a small anisotropy towards the galactic plane, with a rather steep energy spectrum. An estimate of the luminosity function of the medium X-ray sources (Dilworth et al., 1973) shows that at least half this background can be accounted for by unresolved sources of the same type. The remainder could represent the integrated effect of unresolved sources of much lower luminosity than those so far observed, or be due to a true diffuse mechanism.

C. Hard X-rays and low energy gamma rays

A considerable amount of effort has been spent in measuring the diffuse radiation in this region (for a summary see Yash Pal, 1973). Although little has been done to determine the degree of isotropy, the emission is in general ascribed to a predominantly extragalactic background, in analogy with the case of the medium X-rays, given that the relative spectral curves join rather smoothly.

HIGH ENERGY GAMMA RAYS

Apart from studies of the Crab nebula, the observations (Kraushaar et al., 1972) of the OSO-3 satellite of a strong gamma ray emission in the direction of the galactic plane, concentrated within the plane towards the galactic center, remains the cardinal point of gamma ray astronomy to date.

In contradistinction to what occurs in the lower energy ranges, the galactic emission of gamma rays of energy $>$ 100 keV is notably higher than the isotropic extragalactic emission (I_B=(3.0±0.9)10^{-5}/cm^2sec sterad) as shown by the distribution in galactic latitude of the observed counting-rate (Fig. 8). The distribution in

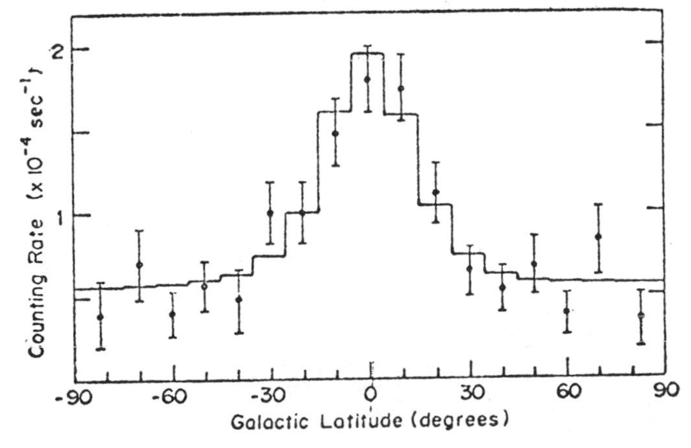

Fig. 8: (Kraushaar et al., 1972) Galactic latitude distribution of all sky events exclusive of those in the galactic longitude range, 330°<l<30°.

galactic longitude (Fig. 9) shows a strong peak in the region of the galactic center ($330^o < l < 30^o$). The authors point out that the angular width of this peak is significantly larger than the angular resolution of the detector (24° FWHM), so it cannot be attributed to a single concentrated source.

The problem again arises as to whether the observed flux of gamma rays is due to diffuse emission or to the integrated emission of unresolved sources. In the first case, Kraushaar et al. (1972) estimate (in the region of the galaxy outside the central peak) the production rate of gamma rays per atom of neutral hydrogen to be $(1.6 \pm 0.2) 10^{-25}$ sec^{-1}. Such a rate can be accounted for within a factor 2, by the decay of π^o-mesons produced in the interaction of cosmic rays with interstellar hydrogen, assuming the same cosmic ray flux as observed at earth. This process cannot explain, however, the emission from the galactic center unless the cosmic ray flux is considerably increased in this region. Ginzburg and Khazan (1972), in discussing this possibility, point out that taking into account the radio flux from this region, one would require also that the ratio of the energy density of the nucleonic component of cosmic rays to that of electrons should be much greater than that at earth, and that equipartition of magnetic field energy and cosmic ray energy could not be maintained.

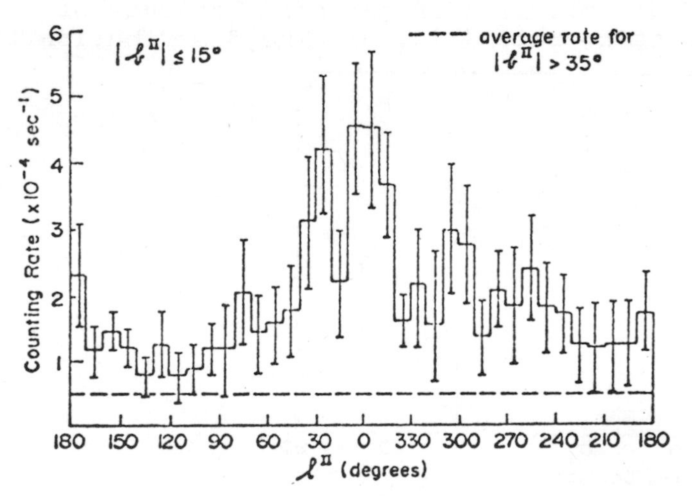

Fig. 9: (Kraushaar et al., 1972) Variation with galactic longitude of the counting rate of sky events in the latitude interval $-15^o < b < 15^o$.

It is difficult to estimate what could be the contribution of unresolved discrete sources, since, at the time of writing, no unequivocal evidence on such sources is available. A very rough estimate can be made considering the relative luminosity of the galaxy in X- and gamma rays. X-ray emission in the medium energy range (2-6 keV) corresponds to a galactic luminosity of $\sim 2 \times 10^{39}$ ergs/sec, while in gamma rays of E>100 MeV, it is $\sim 3 \times 10^{38}$ ergs/sec. If one should assume that the majority of X-ray sources present hard tails to their spectra above $\sim$ 20 keV such as those observed in SCO X1, Cyg X-1, Cyg X-2 and Cyg X-3 (Haymes et al., 1972; Peterson, 1973), and that this tail reaches to the 100 MeV gamma ray region with a spectrum of the type $I(E) = kE^{-\alpha}$ one obtains a gamma ray luminosity of $\sim 10^{38}$ ergs/sec, for $\alpha \sim 2.3$ and a pronounced peak in the longitude distribution in the central region.

Thus, at the present stage it is not possible to determine what fraction of the high energy gamma ray emission is diffuse and how much is due to unresolved sources. It is most probable that a significant part of the emission in the region outside the galactic center is due to cosmic ray produced π^0-mesons, but the problem of the central region must await higher resolution experiments.

REFERENCES

Bell, K.L., Kingston, A.E., M.N.R.A.S. 136, 241, 1967.

Bleach, R.D., Boldt, E.A., Holt, S.S., Schwartz, D.A., Serlemitsos, P.J., Astrophys.J. 174, L101, 1972.

Bleeker, J.A.M., Deerenberg, A.J.M., Yamashita, K., Hayakawa, S., Tanaka, Y., Astrophys.J. 178, 377,1972.

Brown, R.L., Gould, R.J., Phys.Rev. D1, 2252, 1970.

Brown, R.L., Astrophys.Space Sc. 18, 329, 1972.

Cavallo, G., Gould, R.J., Il Nuovo Cimento 2B 1,77,1971.

Chodil, G., Mark, H., Rodrigues, R., Seward, F.D., Swift, C.D., Turiel, I., Hiltner, W.A., Wallerstein, G., Mannery, E.J., Astrophys.J. 154, 645, 1968.

Collet, J., Dilworth, C., Pacault, R., Pounds, K., Rocchia, R., Roth, E.R., Sandford, P., Nature 228, 756, 1970.

Cooke, B.A., Griffiths, R.E., Pounds, K.A., Nature 224, 134, 1969.

Dilworth, C., Maraschi, L., Reina, C., submitted to Astron.Astrophys., 1973.

Friedman, H., Fritz, G., Shulman, S.D., Henry, R.C., IAU Symp. No. 55 (D. Reidel), 215, 1973.

Giacconi, R., Rossi, B., J.Geophys.Res. 65, 2, 1960.

Giacconi, R., Muttay, S., Gursky, H., Kellogg, E., Schreier, E., Tananbaum, H., Astrophys.J. 178, 281, 1972.

Ginzburg, V.L., Syrovatskii, S.I., The Origin of Cosmic Rays, Pergamon Press, 1964.

Ginzburg, V.L., Khazan, Y.M., Astrophys.J. 12, L155,1972.

Gorenstein, P., Tucker, W.H., Astrophys.J. 176, 333,1972.

Gursky, H., Les Houches Summer School on Black Holes, Gordon and Breach, 1972.

Hayakawa, S., IAU Symp. No. 55 (D. Reidel), 235, 1973.

Haymes, R.C., Harnded Jr., F.R., Johnson III, W.N., Prichard, H.M., Bosch, H.E., Astrophys.J. 172, L47, 1972.

Henry, R.C., Fritz, G., Meekins, J.F., Chubb, T.A., Friedman, H., Astrophys.J. 174, 389, 1972.

Hunt, G.C., Sciama, W.D., M.N.R.A.S. 157, 335, 1972.

Ilovaisky, S.A., Astron.Astrophys. 11, 136, 1971.

Kraushaar, W.L., Clark, G.W., Garmire, G.P., Borken, R., Higbie, P., Leong, C., Thorsos, T., Astrophys.J. 177, 341, 1972.

Kellogg, E., Gursky, H., Murray, S., Tananbaum, H., Giacconi, R., Astrophys.J. 169, L99, 1971.

Lilley, A.E., Astrophys.J. 121, 559, 1955.

Martin, P.G., M.N.R.A.S. 149, 221, 1970.

Matilsky, T., Gursky, H., Kellogg, E., Tananbaum, H., Murray, S., Giacconi, R., preprint A.S.E. 3092, 1972.

McCammon, D., Bunner, A.N., Coleman, P.L., Kraushaar,W.L., Astrophys.J. 168, L33, 1971.

Peterson, L.E., IAU Symp. No. 55 (D. Reidel), 51, 1973.

Reina, C., Tarenghi, M., Astron.Astrophys., in press,1973.

Salpeter, E.E., IAU Symp. No. 55 (D. Reidel), 135, 1973.

Seward, F.D., Burginyon, G.A., Grader, R.J., Itill, R.W., Palmieri, T.M., Astrophys.J. 178, 131, 1972.

Stecker, F.W., Astrophys.Space Sc. 6, 377, 1970.

Tananbaum, H.D., IAU Symp. No. 55 (D. Reidel), 9, 1973.

Trümper, J., Schönfelder, V., preprint, 1973.

Yash Pal, IAU Symp. No. 55 (D. Reidel), 280, 1973.

H II REGIONS

H.J. Habing

Sterrewacht te Leiden
Leiden, The Netherlands

I. INTRODUCTION

The study of HII regions is very old. In the 17th century Christiaan Huygens had already observed and described the Orion nebula. Our understanding of this and similar objects increased dramatically between 1925 and 1940 when quantum theory made it possible to interprete the emission line spectra quantitatively. Gradually improving optical techniques have lead to quantitatively reliable observations (especially in the last 10 years). Since about 1960, important new information has come from radio observations in the cm-wavelength range; from the detection of molecular emission in 1963 and from near and far infrared observations since about 1965. Because of these developments we can expect that HII regions will remain at the focus of interest in the coming years.

HII regions are of fundamental importance to astronomy for at least four reasons:

1. They are very short living phenomena and their study may yield information on the problems of star formation.

2. They trace spiral arms and a study of their properties as a function of their location in a spiral arm may yield insight into the formation and maintenance of spiral arms.

3. They enable the determination of elemental abundances and of variations of relative abundances with position in the galaxy.

4. They may provide much or all of the kinetic energy of the (neutral) interstellar gas; they are thus of considerable interest in connection with small scale gas-dynamic phenomena.

In this review I shall attempt to give a broad picture of recent developments in the study of HII regions. I shall deal mainly with aspects 1 and 2, since I expect that these will become better understood in the next few years. However, beautiful observations concerning aspects 3 and 4 have also been made in recent years and these will be briefly mentioned in section VI.

II. GROUP PROPERTIES OF HII REGIONS

1. Definitions

The term "HII region" seems to have been introduced by Strömgren in 1939, when he proved the theorem that a hot star will ionize completely any surrounding gas out to a certain sharply defined boundary. Beyond that boundary the gas suddenly becomes completely neutral (for a concise proof of this theorem see Petrosian et al., 1972, or Strömgren's (1939) - very readable! - paper). If the surrounding gas has uniform density n (actually a very rare situation) this leads to the formation of a sphere of ionized gas, a so-called "Strömgren sphere". "HII regions"+ are usually identified with "Strömgren spheres".

The radius R of the sphere is determined by the equilibrium condition that inside the sphere the total number of recombinations per second equal the total number of ionizing photons per second produced by the star. Since the recombination rate per cm^3 per sec is proportional to n^2, and the total volume of the sphere is proportional to R^3, it follows that the total number of recombinations is proportional to n^2R^3, and this then should be a constant for any given type of star (see section II.5.). Since also the volume emission coefficient is proportional to n^2, the surface brightness is propor-

+Note that this term is unfortunate. "HII" refers to the (non-existent) spectrum of H^+. A better term would have been "H^+-region". The observed spectrum is that of HI.

tional to n^2R, and thus for a given type of star proportional to R^{-2}. It follows that the smallest nebulae are the brightest, but obviously HII regions of any density and any size can in principle be found. (In his introductory lectures van de Hulst warned us of the wild variety of conditions in the interstellar medium!) On the one hand there exist well defined nebulae such as the Trifid, the Omega, the Rosette nebulae and the great Nebula in Orion, on the other hand we find faint extended emission complexes, for instance in the Cygnus region. Recently radio observations have even revealed the existence of widespread, diffuse ionized gas in the central parts of the galaxy.

Infrared observations indicate that the ionized gas contains dust particles, which will also absorb Lyman continuum photons. Thus, less photons are available to ionize the gas and, for a given type of star, the Strömgren radius will shrink. For a quantitative discussion see Petrosian et al. (1972).

In these lecture notes I shall mainly be concerned with the smaller, brighter HII regions. It is important to distinguish them from two, to some extent related objects: reflection nebulae and planetary nebulae. Reflection nebulae consist of dust that scatters back the light of a nearby, very luminous star (types B2 to early A). It is only when a star becomes hotter than B2 that it starts to ionize its surrounndings to an appreciable extent. However, as we will see, even in HII regions like the Orion nebula, the continuum light is largely scattered stellar light, and the transition from reflection nebulae to HII regions is not sharply defined. Since HII regions are dominated by $H\alpha$ at 6500 Å, they appear to be rather red. Reflection nebulae are blue, since they scatter light of rather blue stars. Inspection of the blue and red Palomar Sky Survey plates can often aid in distinguishing reflection nebulae from HII regions. Other objects, somewhat similar to HII regions are the planetary nebulae. Genetically they are quite different: low mass nebulae (0.1 $M_\odot$) arounnd the very hot degenerate core of an evolved star. Still, a planetary nebula can look quite similar to an HII region, and likewise small bright HII regions may be mistaken for planetary nebulae. A well-known example is K3-50 (Rubin and Turner, 1971). In our present context the significance of planetary nebulae is that many of the processes operating inside them also operate in HII regions.

2. Catalogues

Catalogues of HII regions are generally based on observations in one wavelength range: optical, radio or infrared. Each type of catalogue has its own peculiarity. Optical catalogues give objects with very small scale structure (down to 1 arcsec). However, HII regions further away than one or two kiloparsec are excluded because of foreground extinction. Radio catalogues are not limited by extinction and are therefore much more complete. They cover now about one half of the galactic plane. In addition, the data they contain (surface brightness, flux) are quantitative. Infrared surveys single out HII regions where much interaction exists between the stellar UV output and dust, presumably indicating that the HII region is young. They too are not limited by extinction.

As for optical catalogues, the HII regions of the northern sky between $l = 15^{\circ}$ and $l = 240^{\circ}$ are listed in Sharpless' (1959) publication. His numbering system (e.g. S 157) is often used. Note that Sharpless lists the galactic position of the region in the old (l^{I}, b^{I}) system. For a quick conversion to the new (l^{II}, b^{II}) system of galactic coordinates one should use the rule: $l^{II} = l^{I} + 32^{\circ}.4$; $b^{II} = b^{I}$ (b^{II} may be off by as much as $1^{\circ}.5$). For the southern sky the RCW designation is often found, referring to the catalogue by Rodgers et al. (1960).

Radio catalogues are listed in Table 1. It contains only recent catalogues made with beamwidths of about 10 arcmin or less. The sensitivity limit is of the order of a few flux units (1 f.u. = 10^{-26} W.m^{-2}.Hz^{-1}). It is remarkable that the northern sky from about $l = 50^{\circ}$ to $l = 180^{\circ}$ has only been poorly covered. This region contains a large section of the Perseus arm and is certainly very interesting.

A very important step forward has been the publication of a far infrared survey by Frederick et al. (1971). Their bandwidth extended from 80μ to 135μ and their beam was 12 arcmin. Their sensitivity was limited to 10^{4} f.u. The longitude interval is $335^{\circ} < l < 88^{\circ}$.

3. Catalogues of observations in the emission lines

Extensive measurements of radial velocities from the Hα line have been made by the Marseille group. A comprehensive publication with numerous references is given by

Interval in longitude	Interval in latitude	Frequency (GHz)	Wavelength (cm)	Beam (arcmin)	Reference
from 335° to 75°	from -4° to +4°	1.4	21	11	Altenhoff, Downes, Goad, Maxwell, Rinehart (1970)
idem	idem	2.7	11	11	
idem	idem	5.0	6	11	
from 288° to 47°	from -2° to +2°	2.7	6	8.2	Day, Caswell, Cooke (1972) and references contained therein

Table 1a: Catalogues of radio continuum sources in the galactic plane.

from 265° to 50°	from -2° to +2°	5.0	6	4	Goss, Shaver (1970)
idem	idem	0.4	75	3	
from 4° to 235°	mainly from -4° to +4°	1.4	21	10	Felli, Churchwell (1972)

Table 1b: Catalogues of individually selected sources.

Georgelin and Georgelin (1970). It covers the whole sky.

Radio recombination line surveys have been made by Reifenstein et al. (1970) for the northern sky ($348^{o} < l < 85^{o}$) and by Wilson et al. (1970) for the southern sky ($260^{o} < l < 15^{o}$). Besides radial velocities the radio catalogues give the total intensities in the lines and the widths of the lines. The two papers also contain crude estimates of physical parameters derived from these data.

4. Population characteristics

Observations of other galaxies prove beyond doubt that bright HII regions are good tracers of spiral arms. See, for example, the Hα photographs of M 101 (Fig. 1). However, it should be realized that a sizeable fraction of HII regions occur in between the spiral arms. In our galaxy the distribution of HII regions is obtained less directly, but all observations indicate that no important differences exist with other spiral galaxies. Bright HII regions are extreme population I objects, distributed in a flat disk with a strong concentration in spiral arms. It therefore follows that a statistically quite good distance can be obtained from the radial velocity. This is important for the many HII regions that because of their large distance are optically invisible and are detected only at radio and infrared wavelengths.

In recent years it has been found that quite a few "giant" HII regions exist (Reifenstein et al., 1970), defined as HII regions with a value of $n_e^2R^3$ that is four times larger than that of Orion. Probably giant HII regions are the extreme examples out of a continuous size range of HII regions, although this point is not yet very clear. Both observations inside our galaxy (Reifenstein et al., 1970) and in other galaxies (Israel and van der Kruit, 1973) prove that giant HII regions occur in the inner part of a galaxy. The explanation of this phenomenon has very probably to do with the formation of spiral arms and the occurrence of spiral shock waves.

5. Exciting stars

A fundamental requirement from our definition of HII regions is that they are ionized by hot, luminous, i.e. O and early B type stars. Are such stars always found in HII regions? What can be learned from and about them?

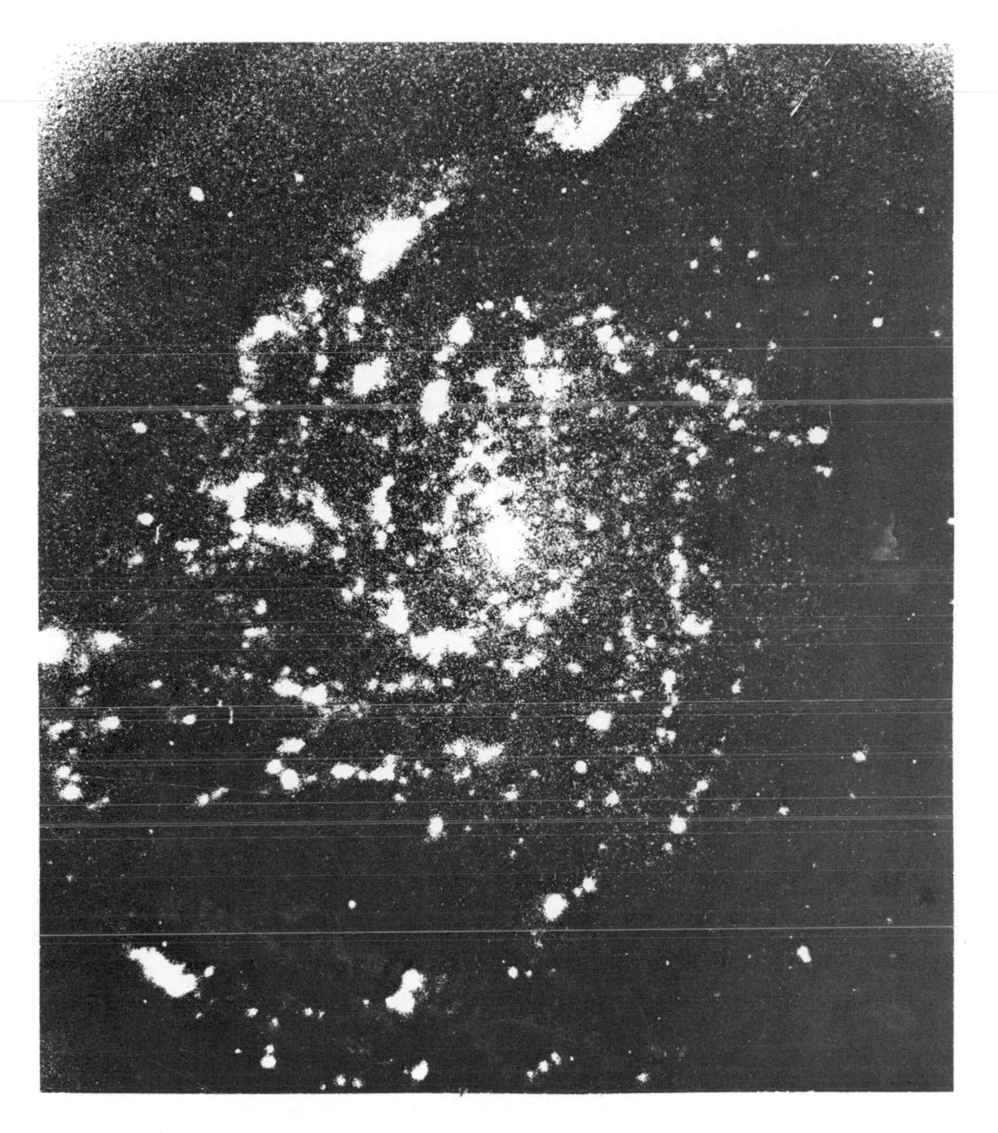

Fig. 1: Distribution of HII regions in the galaxy M 101. Notice that the HII regions follow closely the spiral pattern. (Small band filter photograph at Hα made at the Haute Provence Observatory, France)

Obviously one can only expect to detect the exciting stars in optically visible HII regions; but even there the stars will be quite faint, if the HII region is two or more kiloparsec away. In that case additional arguments are required to identify the correct star. Such arguments may be (Courtès et al., 1968):

1. Location of the star in the HII region (in the center? at the edge?).

2. The star is at the position indicated by bright rims.

3. The star is at a photometric distance agreeing with, for example, the kinematic distance of the HII region.

In many HII regions the exciting stars are known unambiguously. This is the case for most bright, nearby HII regions, with the Orion nebula as a prime example. An extensive list of exciting stars is given by Georgelin and Georgelin (1970). M.C. Lortet (unpublished) has recently identified many OB-stars in the Hamburg/Cleveland catalogue of luminous stars as exciting stars of HII regions. Still, many HII regions exist (especially the most distant ones), where no exciting star has so far been found. Notorious examples are M17, NGC 6334, and NGC 6357.

If the exciting star(s) of an HII region is (are) known, then one can use the photometric distance to obtain a distance estimate apart from the kinematic distance derived from radial velocity. In this way one can check the relation between radial velocity and distance (Georgelin and Georgelin, 1970). The result is valid, however, only if the photometric distance has not been used as a criterion to identify the exciting star.

To determine a photometric distance, one has to correct the observed magnitudes for extinction. Although for several HII regions such a correction can be made by the standard method, using $R = A_V/E_{B-V} = 3$, at least three HII regions give problems: the Orion nebula, M 8 and M 20 show too little extinction in the red compared to predictions (Anderson, 1970). It may be that the interstellar grains in the direction of these three nebulae are systematically larger, but it is also possible that the stars have circumstellar shells giving rise to an infrared excess. Such an excess has recently been measured by Ney et al. (1973) between 2 and 20 μ in the Trapezium stars in Orion.

Finally, let us discuss briefly the parameter S_∞, the total number of photons beyond the Lyman limit, emitted by a star per second. It is clear that S_∞ will not be smaller than the total number of recombinations inside the HII region, so that we have

$$S_\infty > \alpha (T_e) \iiint_{volume} n_e^2 \, dxdydz,$$

where α (T_e) is the rate coefficient for recombination to all levels in hydrogen except the first level. In the same way we can write for the radio flux

$$F_\nu \text{ (radio)} = \varepsilon (\nu, T_e) \iiint_{volume} n_e^2 \, dxdydz,$$

if the frequency is so high that the HII region is transparent. Here $\varepsilon(\nu, T_e)$ is a well-known function of ν and T_e. It follows that

$$S \geq (\alpha/\varepsilon) \, F_\nu \text{ (radio)},$$

so that, if T_e is known, a lower limit for S_∞ can be obtained. Since $(\alpha/\varepsilon) \propto T_e^{-0.2}$ an approximate value for T_e is sufficient. For some HII regions it can be safely assumed that all ionizing photons are absorbed by the gas, and in such cases $S_\infty = (\alpha/\varepsilon) \, F_\nu$ (radio).

Values of S_∞ obtained in this way form a check on values of S_∞ predicted by means of model atmosphere calculations. For a recent discussion, plus a tabulation of S_∞ values, see Churchwell and Walmsley (1972). Instead of S_∞ one often uses the excitation parameter u, defined by the relation

$$u^3 = \iiint n_e^2 \, dxdydz.$$

It follows that $u \propto S_\infty^{1/3}$. If n_e were 1 cm^{-3} then u^3 were the volume of the HII region and u the size of the Strömgren sphere. u then corresponds to, what was called in the older literature, the "Strömgren-radius".

III. THE SPECTRUM OF HII REGIONS

The observed spectrum of HII regions extends from the visible through the infrared into the radio wavelength range. Certainly we will have in the near future also observations in the rocket UV. For extensive lists of optical emission lines observed in the Orion nebula we refer the reader to Kaler et al. (1965) and to Morgan (1971). In what follows we will assume that HII regions are completely transparent to all radiation in the

optical and the infrared wavelength ranges. This is a very good approximation (with one important exception: the Lyman lines). In the radio wavelength range self-absorption does occur. We will mention a few effects in this section.

The following emission processes operate in HII regions.

1. Recombination of ions and electrons. Inside HII regions there is a balance between photoionization and (mainly radiative) recombination of ions. In the recombination process both continua and lines are produced.

2. Collisional excitation. All ions are essentially in their ground state since radiative decay is very much faster than collisional excitation. Lines may be emitted from levels excited by collisions. The strongest lines are emitted in forbidden transitions between low-lying levels but very weak, permitted transitions have been seen in long-exposed spectra of the Orion nebula.

3. Radiation by dust grains. Two forms: firstly scattering of light by dust grains leads to a continuum notably in the visible. Secondly, absorption of light leads to reemission at other wavelengths (notably in the infrared).

We consider these processes now in some detail.

1. Recombination of ions and electrons

<u>(a) Recombination continua</u>. Since we will consider only recombination in binary collisions all rates per cm^3 are proportional to $n_e n_p$, with n_e = electron density and n_p = proton density. In interstellar matter one has always $n_p < n_e < 1.20\ n_p$, if H is ionized by more than 1%. We will assume throughout that $n_e = n_p$ and therefore in all rates (recombination rate, free-free emission rate, free-bound emission rate) a factor n_e^2 will appear. Also the relative velocity v between electrons and protons plays a role. In the expressions for the rates, terms like v, v^{-1}, etc. will occur. Under interstellar conditions the electrons and ions relax almost instantaneously towards a Maxwell distribution characterized by an "electron temperature" T_e (Bohm and Aller, 1947) and instead of v, v^{-1}, one finds $T_e^{\frac{1}{2}}$, $T_e^{-\frac{1}{2}}$. Quite generally a reaction rate may be written as $n_e^2\ f\ (n_e, T_e)$.

All binary recombination rates involving photon emission between electrons and positive ions (point charges of charge + Ze) can be derived from one fundamental expression. See, for example, Kaplan and Pikelner (1970). As such an expression one can use that for the amount of radiation emitted by a gas of positive point charges of charge + Ze (density n_p) and of electrons (density n_e) travelling in random directions with the same speed v. The radiation emitted in ergs per second per cm^3 per Hz is given by

$$\varepsilon(\nu) = \frac{32\pi^2 Z e^6}{3\sqrt{3}\, m^2 c^3 v}\, n_e n_p\, g(\eta_i, \eta_f)$$

m is the electron mass and g is the Gaunt factor, the quantum mechanical correction to the classical equation derived first by Kramers in 1923. η_i and η_f refer to the energy of the initial state (E_i) and of the final state (E_f), respectively., by the relation $E=Z^2\chi_o/\eta^2$; χ_o is the ionization potential of H; both η_i and η_f are real numbers. By admitting that η_f can be imaginary the expression also covers free-bound emission. A closed, but rather complicated expression exists for g, derived in 1935 by Sommerfeld. A thorough discussion of various approximations to g has been given by Brussaard and van de Hulst (1962). The reader is referred to this paper for accurate approximations in any desired wavelength range, for the free-free (ε_{ff}) and free-bound (ε_{fb}) emission coefficients. From these approximations it follows that in the radio frequency range $\varepsilon_{ff} \gg \varepsilon_{fb}$ but that in the visible part of the spectrum $\varepsilon_{fb} \gg \varepsilon_{ff}$. In the radio frequency range only free-free emission contributes to the continuum.

(b) Recombination lines. Following recombination an atom (here: always the H-atom) cascades down from one excited level ($n + \Delta n$) to another (n). The frequency of the line is given by the Bohr equation

$$h\nu_n = \chi_o\left[n^{-2} - (n + \Delta n)^{-2}\right].$$

The lines are numbered by the value of n followed by α, β, γ, etc. for Δn= 1, 2, 3, etc. Thus, H 109α corresponds to the transition from n = 110 to n = 109 (in hydrogen), and H 2α corresponds to the transition from n = 3 to n = 2 (this is the Balmer alpha line, called by an older convention Hα).

Recombination lines divide quite naturally in low-n

transitions and high-n transitions. Low-n transitions ($n < 4$) have been observed in emission in the visible and the near infrared for almost 100 years. High-n transitions ($n > 80$) were first detected in the microwave range in 1963. Lines in the infrared have so far not been detected, but presumably only because of instrumental limitations. Although for both high- and low-n lines, the same basic physical processes operate, differences do occur in the emission mechanism.

To avoid confusion with the usual symbol, n, for density, the symbol N_n will represent the number of H-atoms in level n per cm^3. I assume that N_n is constant in time. For the emission coefficient one then obtains

$$\varepsilon(n+\Delta n,n)=h\nu\Big[N_{n+\Delta n}\left\{A_{n+\Delta n,n}+B_{n+\Delta n,n}\;U\;(h\nu)\right\} - N_n\;B_{n,n+\Delta n}U\;(h\nu)\Big]$$

where A and B are Einstein coefficients. For low-n, stimulated emission and absorption can be ignored (except absorption from n=1) and so $\varepsilon = h\nu\, N_{n+\Delta n}A$ for all visible lines (but not for the Lyman lines!). For high-n the full equation has to be used; it is possible that weak masering occurs (Brocklehurst and Seaton, 1972). To obtain ε one has to calculate N_n. Such a calculation was first done by Menzel in the 1930's. The calculation is done in the following way. For each level n one writes down an equation representing the equilibrium between population processes into, and depopulation processes out of, that level. Population into the level occurs by spontaneous emission and collisional deexcitation from higher levels, by collisional excitation from lower levels and by radiative recombination from the continuum. Depopulation occurs by collisional processes (to lower levels, to higher levels, to the continuum) and by spontaneous emission into lower levels. Since each term contains one factor N_n (with n=1,2,3,...) the equation is linear in N. However, it is inhomogeneous, since radiative recombination leads to a term $N_eN_p\alpha_n$. In total one obtains an infinite set of inhomogeneous linear equations where the inhomogeneous term always contains the factor N_eN_p. By approximative methods of solution, accurate estimates of N_n may be obtained.

The main results can be summarized as follows.
1. The occurrence of a systematic factor N_eN_p in the inhomogeneous term leads to the general result (already mentioned before) that one can write

$$N_n = N_e N_p f_n (N_e, T_e)$$

This equation suggests that it is convenient to use, instead of N_n or f_n, $b_n = N_n / N_n^*$, where

$$N_n^* = \left(\frac{h^2}{2\pi m k T_e}\right)^{3/2} n^2 \exp(X_n / kT_e)\, N_e N_p$$

is the population of the n-th level under conditions of L.T.E. As a general rule $b_n \longrightarrow 1$ as $n \longrightarrow \infty$. From what has been said before it follows that b_n is a function of N_e and T_e.

It turns out that, for low n, b_n is almost independent of N_e, mainly because collisions do not populate the relevant levels. However, for high n, collisional population and depopulation are important and b_n does depend strongly on N_e. The most up-to-date calculations of b_n are those by Pengelly (1964; $n < 20$) and by Brocklehurst (1970; $n > 40$).

2. Since the emissivity in the Balmer-lines (n=2) is given by the relation

$$\varepsilon_{j2} \propto N_e N_p f_{j2} (T_e)$$

it follows that the <u>ratio</u> of two Balmer lines does not depend on $N_e N_p$ and only (but very weakly) on T_e. The same is true for the ratio of any individual Balmer line to the recombination continuum. Therefore the fluxes from an HII region in the various Balmer lines should be in constant proportion to one another and to the continuum. This statement is affected, however, by considering dust extinction. Since dust extinction increases with decreasing wavelength, the Hβ intensity is more weakened than the Hα intensity, the Hγ intensity more than the Hβ intensity, etc. The ratios Hγ/Hβ, Hβ/Hα, Hα/(radio-continuum) therefore decrease with increasing dust extinction. In principle, this offers the possibility of deducing the extinction from the observed ratios. However, the interpretation is simple only if the dust is in front of the HII region, but becomes complicated if the dust is mixed with the ionized gas. For a reliable discussion of the problem see Münch and Person's (1971) interpretation of Hγ/Hβ and Hβ/Hα observations in the Orion nebula, and Mathis' (1970) model calculation.

3. For low-n the width of the lines is determined by the velocity distribution of the atoms, i.e. we have pure Doppler broadening. However, for high-n (generally $n>150$) collisional cross-sections become so large that collisional broadening may dominate and will then determine the line shape. Ultimately, when n becomes very high (say $n > 300$) the lines will be so broadened that they can no longer be distinguished from the background. For a detailed and up-to-date discussion see Brocklehurst and Seaton (1972).

4. A significant contribution to the population of all the levels (but especially the lower levels) is made by reabsorption of Lyman-line photons produced by the gas itself. In terms of the available approximative methods for solving the set of equations for b_n, this means that in HII regions we have to take "case B" solutions, which assume a detailed balance in all the Lyman lines, i.e. $A_{n1}N_n + B_{n1}N_n$ u (Lyman-n) $= B_{1n}N_n$ u (Lyman-n).

Finally I want to make a few remarks on the effects of finite optical depth. (i) As has been noted, self-absorption of the Lyman line photons is very important but can be taken into account in a relatively simple way. No other effects of optical depth are important in the visible spectrum: HII regions are transparent apart from dust extinction. (ii) In the radio regime optical depth effects are important for the continuum flux. The free-free absorption coefficient becomes very large at long wavelengths; taking $T_e=10^4$K we find that the optical depth τ is larger than 1 for $\lambda > \lambda_c$, where $\lambda_c=5\times10^4 E^{-1/2}$cm, and where the so-called "emission-measure" E equals $\int n_e^2 dl$ and is expressed in cm^{-6} pc. If we take $E=5000$ cm^{-6} pc for a normal HII region this means that the region becomes optically thick for $\lambda >750$ cm.

Generally, it is assumed that for small λ, where $\tau<1$, the flux S of an HII region is essentially independent of λ ($S \propto \lambda^{0.1}$), whereas for large λ, where $\tau>1$, $S \propto \lambda^{-2}$. The first statement can be proven rigorously to be true; the second is true only for very special source geometries. The relation $S \propto \lambda^{-2}$ is rigorously valid, for example, if the ionized gas is distributed in a homogeneous sphere. However, if the ionized gas has a spherical density distribution which falls off more slowly, for example, as $n \propto r^{-2}$, then the flux in the optical deep part of the spectrum varies as $S \propto \lambda^{-0.7}$ (Olnon and Habing, 1973). (iii) Optical depth effects are important for high-n recombination lines. Weak masering may even be produced by a rarefied gas in front of a strong radio source (Brocklehurst and Seaton, 1972).

In summary: What can one learn from observations of the recombination continuum and lines? Firstly, as we noted already, the ratios of the Balmer line intensities ($H\gamma/H\beta$, $H\beta/H\alpha$) and the ratio of any individual Balmer line to the (radio) continuum may be used to derive the external dust extinction and, with some uncertainty, the internal dust extinction. A natural limitation arises, however, when so much intrinsic extinction exists that the optical recombination lines are coming only from parts of the HII region closest to the observer, as is probably the case in Orion.

Secondly, from the total flux in the line (or in the continuum) one can derive the integral of n_e^2 over the volume of the HII region. If an estimate of the volume is made one can then derive $\langle n_e^2 \rangle$ and, provided that $\langle n_e \rangle \approx \langle n_e^2 \rangle^{1/2}$ (we will call henceforth $\langle n_e^2 \rangle^{1/2} = n_e$ (r.m.s.)), one can derive the total mass of ionized gas. The resulting mass is rather likely only an upper limit since large density fluctuations are known to exist inside HII regions and therefore $\langle n_e \rangle < n_e$(r.m.s.). Thirdly, from the continuum at wavelengths where $\tau \gg 1$ one can derive the electron temperature (since the surface brightness equals the Planck function), whereas, if the slope of the flux dependence on λ deviates from -2 (e.g. $S \propto \lambda^{-0.7}$), one may obtain some information on the electron distribution in the source. Fourthly, from the high-n recombination lines one may find T_e, provided that masering does not occur. However, regarding masering, the situation is still obscure and we refer the reader to the best available account, that by Brocklehurst and Seaton (1972).

2. Collisional excitation (forbidden lines)

In HII regions the ions and the electrons show a Maxwell velocity distribution corresponding to $T_e \approx 10^4$K, or to an average kinetic energy of about 1 eV. This means that one can expect collisional excitation of ions up to a few eV by electrons in the tail of the velocity distribution. (These excitations will be followed by spontaneous emission: the transitions are usually electric dipole forbidden.) Neither hydrogen nor helium possesses levels of such low excitation potential, but ions with their valence electrons in p and d shells do. The excited levels derive from the interaction between the spin and orbital momenta of the individual electrons. We will restrict ourselves to p-shell ions. First note that a p-shell can contain at most 6 electrons. The configuration p^6 will lead to only one single level; the configurations p and p^5 show an identical multiplet structure, as do p^2 and p^4. This means that we have basically three different types of multiplet structures (see Fig. 2).

configuration p or p^5	configuration p^2 or p^4	configuration p^3
	^1S	^2P
	^1D	^2D
^2P	^3P	^4S
examples: C II (156μ), N III (57μ), O IV (26μ), Ne II (13μ), Si II (35μ)	visible transitions: $^1S \rightarrow {}^1D$, $^1S \rightarrow {}^3P$, $^1D \rightarrow {}^3P$. infrared transitions within the ^3P-level. examples: C I, N II, O I, O III, Ne III, Si I, S I, S III.	visible transitions: $2P \rightarrow 2D$, $2D \rightarrow 4S$, $2P \rightarrow 4S$. no infrared transitions to be expected. examples: N I, O II, S II.

Fig. 2. Multiplet structure for ions with p^q-electronic configurations (q=2,3,4,5).

Since the absorption coefficients are extremely small, stimulated emission or absorption do not play a role. Therefore the emission coefficient is given simply by $\varepsilon = h\nu N_n A_{nn'}$, and the problem is to calculate N_n. To do this, one sets up a set of k linear equations (k=total number of levels, e.g. k=5), where each equation describes the balance between population and depopulation processes for that level. To determine N_n, one solves the set of equations, which is in principle easy; however, it is very difficult to obtain good estimates for the collisional cross-sections, and therefore the transition rates between the levels are often poorly known. For a discussion of the emission coefficients of infrared lines we refer the reader to Petrosian (1970), for a discussion of emission coefficients of visual lines to Seaton (1960), to Eissner et al. (1969), and to Czyzak et al. (1970).

Consider the [OII]-lines λ 3729 Å and λ 3726 Å belonging to the $^2D \longrightarrow {}^4S$ transition of O^+. If N_e is large (in this case: $N_e > 10^4$ cm^{-3}), collisional excitations and deexcitations will balance, and the two ^2D levels will be populated according to LTE conditions, in fact according to their statistical weights (2J+1). Since the upper ^2D level has J=3/2 and the lower J=5/2 the intensity ratio $\varepsilon(\lambda 3726$ Å$)/\varepsilon(\lambda 3729$ Å$)$ will equal (2x3/2+1)/(2x5/2+1) = 2/3. However, if N_e is small ($N_e \ll 10^4$) the population ratio between the two levels will depend only on the collisional excitation rate from the ^4S level and it turns out that then $\varepsilon(\lambda 3726$ Å$)/\varepsilon(\lambda 3729$ Å$) \approx 2.5$. Since the lines are optically thin (the transitions are electric dipole forbidden and the absorption coefficients are therefore very small), the line intensities must have the same ratios as the ε's. From the observed ratio I (λ3726 Å)/I (λ3729 Å) one can therefore interpolate what the true electron density is. A similar argument can be used for the λ 6716 Å, λ 6730 Å doublet for [S II].

Consider also [O III], which corresponds to a $2p^2$-electron configuration. In this case the $^1S \rightarrow {}^1D$ transition corresponds to 4363 Å, the $^1D \rightarrow {}^3P$ transitions to 5007 Å, 4959 Å, and 4933 Å (the last transition has a much lower probability). It turns out that the intensity of λ4363 Å compared to the sum of the intensities of λ 5007 Å and λ 4959 Å is very sensitive to the electron temperature. If one increases the temperature, the ^1S level will be more strongly populated with respect to the ^1D level resulting in an increase of λ4363 Å with respect to λ5007 Å and λ 4959 Å.

We conclude that by observing several lines belonging to the same multiplet the populations of the multiplet-levels can be determined. From these populations one can derive such fundamental properties as the electron density and the electron temperatures. Two difficulties hinder this approach: the knowledge of transition rates is still poor, and extinction affects different lines in different ways. Extinction problems will be decreased for infrared lines. Infrared lines therefore offer many prospects, which the reader will find more extensively discussed by Petrosian (1970).

3. Radiation by dust grains

Dust grains contribute in two ways to the continuum radiation of HII regions. In the first place they scatter light. For example, the continuous spectrum of Orion in the visible is dominated by dust scattered stellar light (see section V.1.). In the second place the grains radiate in the infrared. The energy radiated is balanced by efficient absorption at visual and UV wavelengths. For the emission coefficient one writes rigorously the

Kirchhoff relation

$$\varepsilon(\nu) = k(\nu)\ B_\nu\ (T_g)$$

The fundamental problem is to know the absorption coefficient $k(\nu)$. This depends on the composition of the grain material, which is unknown. Often one writes $k(\nu) = \pi a^2\ Q(\nu)$, where a is a "typical" grain radius and Q is an efficiency factor. A relationship of the form $Q \propto \nu$ is often adopted. Although there is some observational evidence to support this relationship in the near infrared range (Pottasch, 1972) it is not certain that one can extrapolate the relationship into the far infrared. At this institute Andriessen and Olthof suggest

$$Q \propto \nu^p \text{ with } p = 2 \text{ or } 3 \text{ for } \nu < \nu_o \approx 10^{12} \text{ Hz.}$$

It is clear that as long as the nature of the grains is unknown, it is impossible to predict $\varepsilon(\nu)$. Probably the problem should be turned around: we will learn about the nature of the grains, once we have a large number of infrared observations of high spatial and spectral resolution. But this goal is probably still far off.

IV. NEUTRAL MATTER AND DUST ASSOCIATED WITH HII REGIONS

Looking at photographs of HII regions it is immediately evident that HII regions are associated with dark matter. There is ample evidence for dark streaks and objects close to, and possibly inside HII regions. This brings up the question of how much of an HII region will be invisible because of extinction. The answer can be obtained by mapping HII regions at radiowavelengths.

Conclusive evidence that sometimes the major part of an HII region may be hidden from sight due to dust in the immediate vicinity of the HII region came in 1969 from a survey by Schraml and Mezger of HII regions at 2 cm with 2 arcmin resolution. They discovered, for example, that of the HII regions M 17 (= NGC 6618) and IC 1795 (better known as W 3) major parts were completely invisible. On the other hand, the radio and optical pictures coincided more or less in the cases of the Orion nebula (NGC 1976) and NGC 2024 (or Orion B). Clearly HII regions contain "hidden components" and, as it turns out, these components may have the highest surface brightness, and therefore the highest emission measures. Recently, Israel et al. (1973) have shown that the phenomenon also occurs on a smaller scale. For example, S 157, an otherwise inconspicuous HII region around l = 110 contains a very small compact source exactly at a spot where strong local absorption occurs (Fig. 3).

We will pursue the nature of such compact, hidden components in the next section. In the meantime we shall consider the general occurrence of large masses of neutral matter and dust near and inside HII regions.

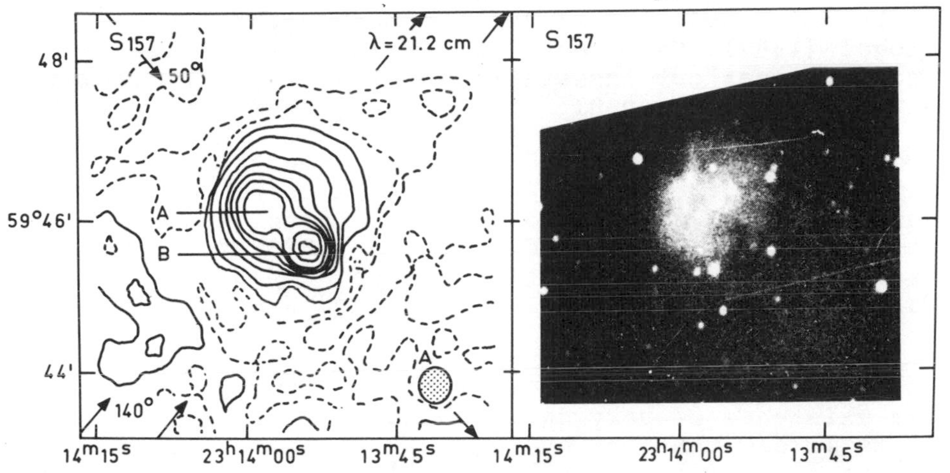

Fig. 3: The HII region S 157, seen at 21 cm wavelength and at Hβ. The radio map (left hand side) contains two components: an extended source (called "A") and a very compact one (called "B"). The optical photograph (right hand side) shows source A clearly, but displays strong extinction at the position of source B. A preliminary radio map obtained at 6 cm with higher resolution shows the same effect in an even more pronounced way. The radio map has been obtained at Westerbork, the Hβ photograph at the Haute Provence Observatory (Chopinet and Lortet-Zuckerman, 1972).

1. The presence of neutral matter

An old question concerning HII regions is the following: does the central star ionize all material directly related to the HII region or does its transition zone occur far outside the original cloud in which the star was born? In the first case the HII region is said to be "ionization-limited", in the second case "density-limited". The question has considerable importance for problems of star formation: how much material is required to build stars, and how efficient is the star formation process?

Consider the possibility of atomic hydrogen associated with HII regions. To investigate this one can make 21-cm line studies. Here the difficulty is to distinguish the HI gas associated with the HII region from the background HI gas, which in galactic 21-cm line work is always strong and irregular. The most complete and convincing study is probably the one by Riegel (1967). He observed 27 small diameter HII regions and got clear cut answers in 9 cases. Two HII regions appeared to have an HI region around them, and 7 definitely had no significant HI associated with them. In the two cases the neutral hydrogen was some 25 times more massive than the ionized hydrogen. A third convincing case of HI gas associated with HII is an enormous cloud found by Bridle and Kesteven (1970) and associated with the group of HII regions known as W 58, which contains as visible objects NGC 6857 and K 3-50.

It may be, however, that there is neutral hydrogen, but in molecular, not in atomic form. Photographs reveal the existence of very dark clouds near HII regions and it is likely that the hydrogen in these has turned into H_2, inspite of the very high radiation density nearby. There is now some good evidence that molecular clouds are indeed associated with HII regions. Probably the best cases are W 49, W 51 and Orion. The last case will be discussed in section V.1. In W 49 and W 51, Scoville and Solomon (1973) have found CO-emission associated both in extent and in radial velocity with the giant HII regions. The authors estimate in both cases that the mass of the neutral cloud is between 10^4 to 10^5 $M_\odot$, that is at least ten times the total amount of ionized gas. There can be no doubt that in the next few years similar clouds will be found near other HII regions.

Summarizing, one may conclude that several HII regions are associated with clouds of neutral matter of considerable mass. Often an HII region appears to be nothing else but a small part of a big cloud, ionized by a star that happened to be born in that part of the cloud. However, not all HII regions can be described in this way. It may well be that smaller objects are in fact density-bound, i.e., the whole cloud has been ionized.

2. Dust inside HII regions

The existence of dark clouds associated with HII regions proves the presence of dust near HII regions. However, we do not know whether dust occurs mixed in with the ionized gas. The first positive evidence to this point was presented by O'Dell and Hubbard (1965), who showed that in the Orion nebula the visible continuum was much stronger and bluer compared to Hβ than could be expected from a recombination continuum. In addition, the ratio (visible continuum/Hβ) increased as the distance to the Trapezium stars increased. They concluded that the visible continuum of the Orion nebula is produced by scattering. Their conclusion was based upon comparing the observed continuum with a continuum predicted in the same way as has been outlined here in section III.1.(a). It is confirmed by the observation of Peimbert and Goldsmith (1972) that the continuum of Orion contains the HII λ 4686 Å line in absorption, and not in emission. Since HeII is not found in the Orion nebula but gives strong lines in the exciting stars, the conclusion is that the continuum is scattered stellar light.

Assuming that the continuum is formed in the same region as the line emission it follows that there is dust inside the ionized region. What are the dust properties? Mathis (1970) made model calculations of dusty HII regions, which show how difficult it is to extract quantitative information from observations such as those made by O'Dell and Hubbard (1965). Probably the best quantitative analysis is that by Münch and Persson (1971) of gas and dust in the Orion nebula. Their analysis is similar to that of Mathis (1970) but much more simplified. They conclude that the reddening of the nebula can be understood, if dust and gas are thoroughly mixed. This conclusion is confirmed by the observation that when the reddening increases, the line intensity also increases. Münch and Persson even suggest that the Hβ radiation we receive is emitted at the first 10 per cent of the line of sight through the nebula - the remaining 90 per cent is obscured by internal dust.

Undoubtedly the best indications for mixing of dust with ionized gas comes from infrared observations. In 1969 Ney and Allen discovered that at 20μ the Orion nebula shows radiation considerably stronger than can be expected from the ionized gas. It is generally assumed that this infrared radiation is emitted by dust grains which are heated by stellar UV radiation. The nebula coincides nicely with the most intense part of the Orion nebula, as observed by Schraml and Mezger (1969) at 2 cm. There is also close agreement of the radio contours with the infrared contours at 10μ and 20μ for M 17 (Kleinmann, 1973; Lemke and Low, 1973); and with the infrared contours of radiation in a band between 40μ and 350μ for NGC 6334 and NGC 6357 (as reported by Emerson at this institute; see Fig. 4). Previously Hoffman et al. (1971)

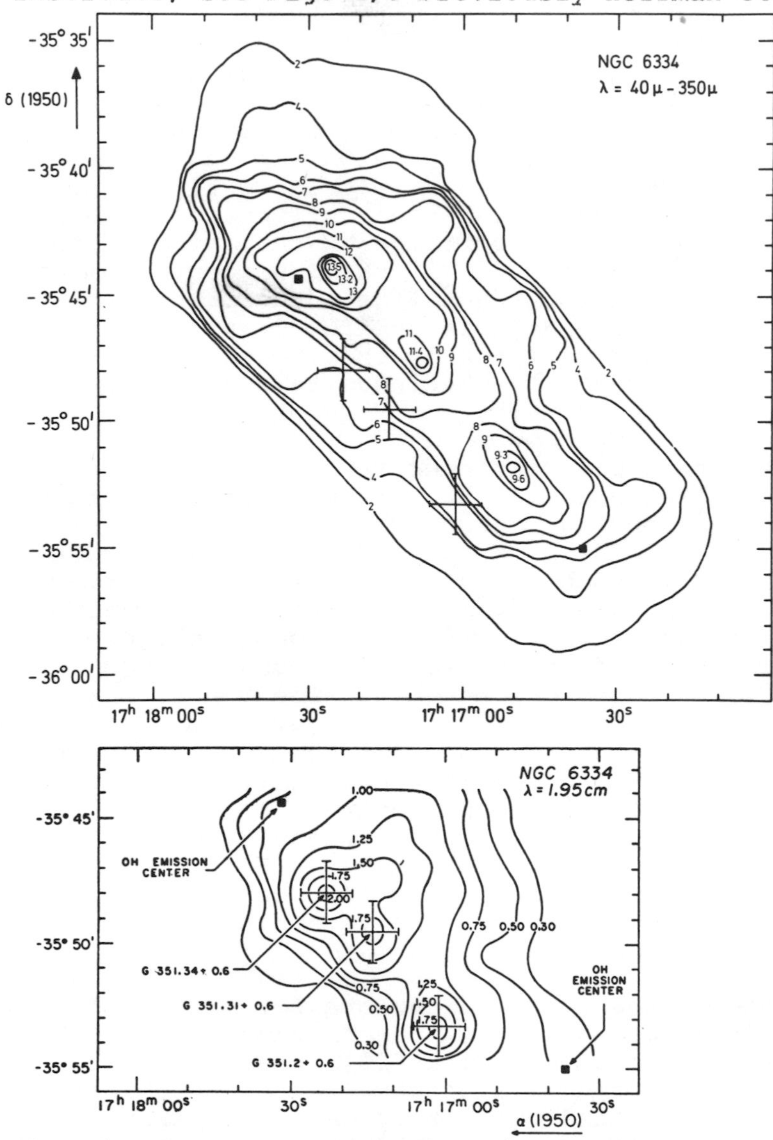

Fig. 4: The HII region NGC 6334 at far infrared wavelengths and at 2 cm. The lower map is due to Schraml and

Mezger (1969). It shows the distribution of the free-free radiation in the nebula at 2 cm wavelength. The upper map is due to the IR-group at University College, London, and was presented at this Institute by Dr. J. Emerson. Notice that the two maps are quite similar. There is some difference in the exact positions, but this is most likely due to technical problems in the infrared positioning. The OH emission centers do not leave any significant mark on the radio map, but the northern OH source has a conspicuous IR counter part. The infrared emission is presumably thermal reradiation by hot dust grains.

---

reported the discovery at 100μ of 72 sources in the galactic plane, of which 46 coincide with HII regions. It thus appears that quite a few HII regions display emission by heated dust grains. However, to date only HII regions with high surface brightnesses at radio wavelengths have been detected at IR wavelengths. This is probably due to the limited sensitivity available for IR-observations.

Two questions of interpretation emerge from the IR-observations: what is the nature of the grains, and how are they heated? Definite answers to these questions do not yet exist. For the total IR-flux S(IR) one can write (if the source is optically thin, which is probably a safe assumption)

$$S(IR) = B_\nu (T_g) Q(\nu) \int n_g dV$$

where T_g is the grain temperature, Q has been defined before, n_g is the grain density and V the volume of the HII region. The nature of the grains could be deduced if, for example, $Q(\nu)$ could be measured. Given the possibility that T_g may vary throughout the source, and that S has been measured in very broad bands at only a few wavelengths, it appears at the moment impossible to separate $Q(\nu)$ from $B(T_g)$ in the observed results. Certainly the future holds great promises.

What heats the grains? Two possibilities are open: either the stellar radiation is first absorbed by the gas and reemitted as radiation at longer wavelengths (mainly Balmer lines, Ly-α lines and forbidden lines) or the stellar radiation is first absorbed by the grains and re-emitted as infrared radiation. If the second possibility is realized the star will ionize less gas than corresponds to its excitation parameter (see, for example, Mathis,

1971 or Petrosian et al., 1972). Therefore the total radio flux from the HII region will be smaller. However, the infrared flux is perhaps not much affected by what kind of photons the grains are heated, so that the ratio of infrared to radio flux may be a sensitive indicator for the exact heating mechanism. For example, the IR-observations of the Orion nebula imply that about 3/4 of the direct stellar light is absorbed by dust grains (Petrosian et al., 1972). Still, it appears to me that the observations leave much uncertainty. (It may well be that the heating mechanism varies from one HII region to another.) More and more detailed IR-observations in the near future will certainly help to resolve this problem.

V. COMPACT AND ULTRA COMPACT HII REGIONS AND THEIR ASSOCIATION WITH OH AND H_2O MASERS

1. The Orion nebula

It is of interest to summarize briefly our knowledge of the Orion nebula (NGC 1976). The most dominant component of the stellar population is the so-called Trapezium, a group of 4 stars, separated from each other by about 10 arcsec.They are called θ^1Ori A,B,C,D. (Some 1.5 arcmin to the southeast one finds two stars called θ^2Ori A and θ^2Ori B.) The brightest star is θ^1Ori C with V=5.16 and spectral type O6p. The other three are B type stars. Within 10 pc from the Trapezium some 18 stars of type B3 and earlier are found, all belonging to the association I c Ori (Blaauw, 1964). The area is also full of very weak emission line stars which are related to T Tauri stars.

If we turn to the nebular emission, we note first that a considerable fraction of the observed continuum is due to scattering by dust particles, mixed with the ionized gas (see section IV.2.). Confirmation that the dust is mixed with the ionized gas is given by the presence of an infrared nebula (the Ney and Allen nebula), which coincides with the brightest part of the radio continuum source and with the Trapezium. The radio continuum observations (Schraml and Mezger, 1969) indicate that the ionized gas is distributed in a spherically symmetric way. This is in part due to beam smoothing since optical observations (see e.g. the Wurm and Rosino atlas, 1959, 1965) indicate that on a small scale (tens of arcseconds) the nebula is very irregular. However, the radio observations should give the best picture of the ionized gas on a scale of 1 arcmin, corresponding to 4.5×10^{17} cm at the distance of Orion. The nebula then appears to consist of a small, very dense central part, surrounded by a nebula of considerably lower density. Representative numbers for the central part are n_e(r.m.s.) $\approx 10^4 cm^{-3}$, diameter d=0.65 pc and emission measure E=2.6 $\times 10^6 pc\ cm^{-6}$.

From the observation of [SII] lines and of [OII] lines electron densities can also be obtained. They lead to values of $n_e \approx 10^5\ cm^{-3}$ at the center of the nebula (Danks and Meaburn, 1971). In general, the forbidden line densities appear to be higher than those derived from the radio observations. If this effect is real, then it is most easily explained by assuming that a part of the ionized gas is concentrated in irregular, dense blobs (Osterbrock and Flather, 1959). These blobs may be neutral in their centers, thus forming "globules". The "globules" will be ionized from the outside and gradually evaporate. The gas streaming from the "globules" may be the explanation (Dyson, 1968a, 1968b) for the complex motions observed in the nebula by Wilson et al. (1959) (see the graphical display of these measurements by Fischel and Feibelman, 1973).

About 1 arcmin north-west of the Trapezium, Kleinmann and Low (1967) discovered another small nebula emitting at 20μ, which in no way corresponds to any feature in the radio continuum. This nebula turned out to be the center of a very rich molecular cloud (see, for example, Solomon, 1973 or Turner, 1973), in its total variety of molecules only surpassed in number by Sgr B2. Since it is difficult to understand the presence of molecules inside an HII region and since the contours of the nebula do not correspond to features in the radio continuum, it is quite suggestive that the Kleinmann-Low (KL) nebula is actually separated from the HII region. Absorption line measurements in the formaldehyde line at 6 cm and at 2 mm suggest that the KL nebula is behind the Orion nebula (Kutner and Thaddeus, 1971). It therefore seems highly likely that the Orion nebula is an ionized fragment on our side of a larger cloud.

Finally, in the center of the IR nebula a point-like (diameter less than a few arcsec) IR-source was discovered by Becklin and Neugebauer (BN) (1967). They ascribe to it a blackbody temperature of 700 K. The object has been interpreted as a heavily obscured supergiant but recent observations at 20μ rule out such an explanation (Wynn-Williams et al., 1973). It is currently thought to be a collapsing protostar. The BN "star" coincides within 1 arcsec (= 500 A.U.) with a strong OH maser (Raimond and Eliasson, 1969). Quite nearby, but at a very definite distance of about 20 arcsec ($=10^4$ A.U.) three H_2O masers are present (Hills et al., 1973).

We conclude that the Orion nebula is a small (<1 pc), dense ($10^5 > n_e > 10^3$ cm^{-3}) HII region with smooth overall structure, but large density fluctuations inside and possible large blobs of neutral gas. Quite near to it, but not coincident there is a hot molecular cloud with at its center an IR star and associated OH maser.

2. Compact HII regions

The central object in the Orion nebula has been called a "compact HII region". In recent years it has become clear that the Orion nebula is not unique, and that several compact HII regions exist, many of which also show associated phenomena such as nearby OH masers, infrared sources or extended molecular clouds.

One of the earlier examples is DR 21, a very small and compact HII region in the Cygnus X complex. According to Wynn-Williams' (1971) high resolution study, its main component has diameter d=0.25 pc and n_e (r.m.s.) $=1.5\times10^4$ cm^{-3}, corresponding to an emission measure $E=n_e^2d=5\times10^7$ cm^{-6} pc. It requires at least an O6 star for tis excitation. At about 3 pc distance a very strong OH maser is found. Both DR 21 and the OH maser (called W 75 S) are strong IR sources.

A very extensively studied area is that of the radio source W 3, adjacent to the visible HII region IC 1795. The main component of the radio source (called W 3 (cont.) or G 133.7+1.2) consists of 4 compact HII regions (Wynn-Williams, 1971). IR-observations have shown (Wynn-Williams et al., 1972) that each of these compact regions also emits between 2 and 20μ, apparently by thermal radiation from heated dust (Fig. 5). In addition, several point-like IR sources have been detected not coinciding with any radio continuum features. For example, the largest of the 4 compact HII regions shows a shell structure. In the center of the shell an IR point source was found at 2μ which is probably a very reddened O type star; no radio continuum is associated with this point source. Another interesting object is IRS 5, an IR point source without radio continuum counterpart, but coinciding

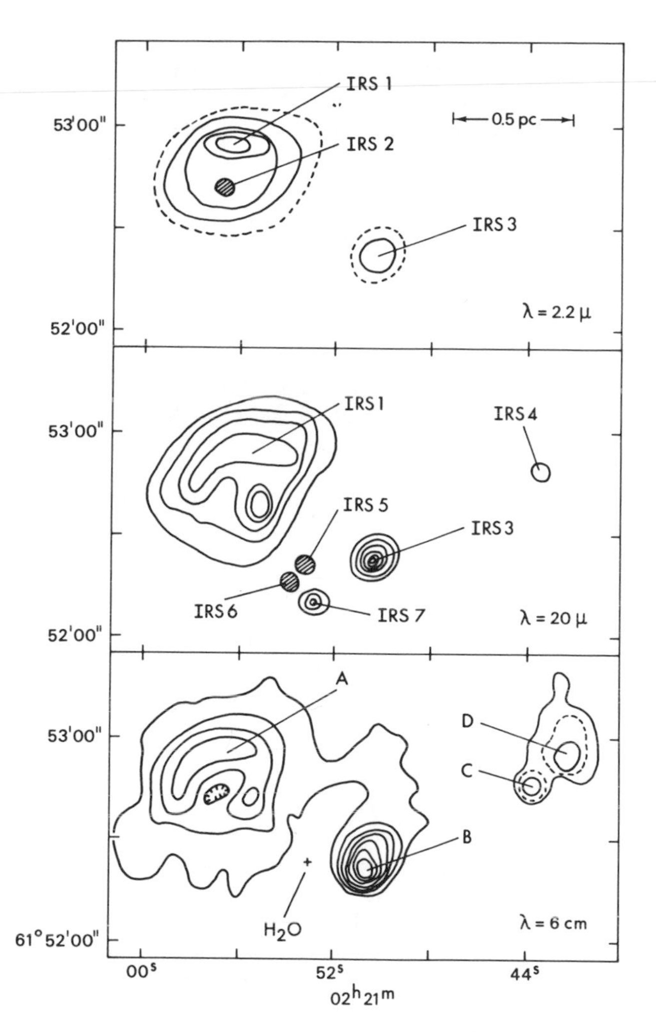

Fig. 5: Isophotes at two IR wavelengths and at 6 cm wavelength (Wynn-Williams et al., 1972). Notice the similarity between the three maps.

within observational errors ($\pm$ 10^4 A.U.) with a strong H_2O maser.

Some 20 arcmin away from W 3 (cont.) is one of the strongest OH masers in the sky. It coincides with a very small HII region which we shall discuss in the next subsection.

Finally, we note that, whilst many compact HII regions (as, for example, the W 3 regions, or DR 21) are optically invisible, others (as for example, Orion A) are very visible. A rather modest example of the optically visible type is S 157 A (or G 111.30-0.66), which can be distinguished as a small (1 arcmin) very bright nebula embedded in a large diffuse nebula, S 157 (Fig. 3). For good photographs and a discussion of optical properties see Chopinet and Lortet-Zuckermann (1972). In the center of the object a late O-type star is found. From measurement of the [SII] doublet line ratio a density $n_e > 600\ cm^{-3}$ has been derived. In the south-east corner of the object a small dark area of less than 10 arcsec across appears superimposed on the nebula. What makes S 157 A extra interesting is that this dark area coincides with a weak, unresolved point source at 21 cm and at 6 cm wavelength (Israel et al., 1973). Apparently this is a very small, very compact HII region with so much dust around it that it is optically invisible. Characteristic values for this object are: diameter < 0.2 pc, n_e (r.m.s.) > 2000 cm^{-3}.

3. Ultra-compact HII regions and maser sources

We remarked in the previous subsection that at about 20 arcmin (or 20 pc) from the complex radio source W 3 (cont.) a very strong OH maser is found. This OH maser, like the ones in Orion and near DR 21 is of a very characteristic type, called class I (dominant in the main lines at 1665 and 1667 MHz, the radiation being strongly polarized, usually circularly). Class I OH masers appear so often, so near to compact HII regions that, although selection effects play a role in the discovery of OH emission sources, the near-coincidence cannot be considered fortuitous (Habing et al., 1972). In the case of the maser W 3 (OH), at 20 arcmin distance from W 3 (cont.) it turns out that the maser coincides, within observational errors (3 arcsec or 0.1 parsec), with an HII region of very small dimensions: 1 arcsec or 2000 A.U. (Wink et al., 1973). The source is optically thick at cm-wavelengths and was discovered at 2 cm. From the radio flux the r.m.s. electron density is estimated to be $2x10^5\ cm^{-3}$. Its excitation therefore requires at least an O8 star.

The radio source W 3 (OH) is probably the prototype of what may be called "ultra-compact" HII regions - objects with typically, say, $n_e \gtrsim 10^4\ cm^{-3}$ and d < 0.1 pc. Within the last year three more such objects have been found: one coicident with the OH maser near NGC 7538

(G 111.5+0.8) (Ryle, 1972); and one coincident with the OH maser ON 2 near G 75.8+0.4. Especially interesting is the coincidence of an optically invisible, ultra-compact HII region (G 69.5-1.0) with an OH maser called ON 1 (Winnberg et al., 1973). This ultra-compact source is the only object visible at 6 cm within an area of about 10 arcmin across. If ultra-compact HII regions form around new stars, then ON 1 may well be the beginning of the formation of a whole new star cluster.

The detection of at least four ultra-compact HII regions leads to the question whether all OH masers of class I coincide with such regions. No definite answer to this question exists. So far, no ultra-compact HII regions have been detected in W 75 S (near DR 21), in W 3 (cont.), or in the Orion nebulae. In all three cases the presence of a nearby, strong radio source makes such a detection very difficult. However, the question posed appears significant. It may well be that the existence of an ultra-compact HII region is related to the mass of the central star and the state of its evolution, for example, whether it has succeeded in reaching the main sequence. The answer to the question may provide the basis for a sub-classification of class I OH sources, with the hope that a sub-classification will narrow down the range in intrinsic properties of these masers.

What interpretation can be given to the ultra-compact HII regions? A very interesting suggestion has been made by Kahn (see his lecture notes in this volume). A massive star, after reaching the main sequence, will still experience a large inflow of material. Ultimately the radiation pressure of the star will stop the inflow. The process of stopping the gas may lead to the build-up of a shell of dense neutral gas, surrounding a sphere of ionized gas. The sphere may correspond to the ultra-compact HII region. The dense shell will eventually break up due to a Rayleigh-Taylor type instability. Inside the fragments OH masers may form. This view is supported by VLBI-observations, which resolve the OH masers into groups of point sources. The model suggested by Kahn poses several new and pertinent questions. Does every OH maser have an ultra-compact HII region associated with it? What are the IR properties of OH maser sources? Do H_2O maser sources have ultra-compact HII regions? (The positions of OH and H_2O masers near the Orion nebula suggest that the physical separation between OH and H_2O masers is sometimes so large that their interaction is very slight.)

I expect that observations of the relation of OH masers to ultra-compact HII regions and to IR sources will be of considerable help in the construction of realistic models to explain these masers. Such models may also help to understand the complicated phenomena accompanying star birth, about which virtually nothing is known. Observations with higher sensitivity revealing still fainter ultra-compact sources and their possible time-variation will be very important.

The relationships between various objects discussed so far in this section have been summarized in Table 2. This table should be considered as a summary of available observational data and has to be used with care. Our knowledge of molecular and infrared emission of matter in and near HII regions is incomplete and highly biased by selection effects and instrumental sensitivity. These may have introduced differences in Table 2 that are physically not realistic.

VI EPILOGUE

In these notes we have emphasized those aspects of HII regions that are currently at the focus of interest: the relation between dust and gas, IR sources and molecular sources. The bias is also a rather personal one.

At least two other aspects could (and actually should) have been included: abundances and kinematics. Regarding abundances it is probably fair to say that any interpretation of spectra in terms of abundance differences or variations is made difficult by the large inhomogeneities that characterize HII regions. Probably the only firm results are from the observation of hydrogen and helium high-n recombination lines by Churchwell and Mezger (1973), which indicate an overall abundance (He/H) of about 8 per cent, except in the central regions of the galaxy, where virtually no helium could be found!

We have also omitted (for lack of time) a discussion of kinematics of HII regions. This subject received much attention in the last decade, when several models were constructed that described the expansion of HII regions around newly formed stars. However, the applicability of such models to real HII regions has been seriously questioned, because the models assume homogeneous gas distributions and rather simple-minded star formation models. For a good review see Mathews and O'Dell (1969).

	Molecular emission	Radio continuum emission	Far IR emission	Near IR emission	Examples
Nebulae I	extended source	none; but HII region may be adjacent	none	none	W 49, W 51; Heiles' clouds?
II	extended sources	none; but HII region may be adjacent	strong	strong optically thick at 20 μ?	Kleinmann-Low nebula
III	none	compact components present	strong	optically thin at 20 μ?	M 17; Orion (Ney and Allen); NGC 6334;NGC6357
Point Sources I	OH maser of classI or H_2O maser	none	?	yes	BN object; W 3-IRS 5; W 3 (cont.)-OH
II	OH maser of classI or H_2O maser	ultra-compact HII region	?	yes	W 3 OH, ON 1, ON 2, NGC 7538

Note: The apparent differences between "Nebulae I" and "Nebulae II" and between "Point Sources I" and "Point Sources II" may reflect instrumental limitations and therefore be either only qualitative or completely fictitious.

Table 2: Summary of IR, molecular and radio continuum properties.

In recent years many excellent observations of the internal kinematics of HII regions have been made by Fabry-Perot techniques, notably by the Marseille group, by Meaburn in Manchester, U.K., and by Weedman and Smith at Kitt-Peak and Cerro-Tololo. However, we lack a reliable interpretation of the "turbulent" velocities found. For example, the excellent observations of the complex internal motions in the Orion nebula (made in 1959!) by Wilson, Münch, Flather and Coffeen have not yet been adequately explained. Probably the best attempt was that by Dyson (1968a, 1968b).

ACKNOWLEDGEMENT

I thank Dr. Anthony Whitworth for his comments and his improvements of the English text.

SUGGESTED READING

Pottasch, S.R., The Diffuse Emission Nebulae, in Vistas in Astronomy, ed. A. Beer, 6, 1965.

Kaplan, S.A., Pikelner, S.B., The Interstellar Medium, Harvard Univ. Press, Cambridge, Mass., 1970.

Terzian, Y. (ed.), Interstellar Ionized Hydrogen, Proc. Symp. Nat.Radio Astron.Obs. Charlottesville, USA, 1967; Benjamin, Inc., New York, 1968.

REFERENCES

Altenhoff, W.J., Downes, D., Goad, L., Maxwell, A., Rinehart, R., Astron.Astrophys.Suppl. 1, 319, 1970.

Anderson, C.M., Ap.J. 160, 507, 1970.

Becklin, E.E., Neugebauer, G., Ap.J. 147, 799, 1967.

Blaauw, A., Ann.Rev.Astron.Astrophys. 2, 213, 1964.

Bohm, D., Aller, L.H., Ap.J. 105, 1, 1947.

Bridle, A.H., Kesteven, M.J.L., Ap.J. 75, 902, 1970.

Brocklehurst, M., M.N.R.A.S. 148, 417, 1970.

Brocklehurst, M., Seaton, M.J., M.N.R.A.S. 157, 179, 1972.

Brussard, P.J., van de Hulst, H.C., Rev.Mod.Phys. 34, 507, 1962.

Chopinet, M., Lortet-Zuckermann, M.C., Astron.Astrophys. 18, 373, 1972.

Churchwell, E., Mezger, P.G., Nature 242, 319, 1973.

Churchwell, E., Walmsley, C.M., Astron.Astrophys. 23, 117, 1973.

Courtès, G., Georgelin, Y., Georgelin, Y., Monnet, G., Pourcelot, A., in Interstellar Ionized Hydrogen, ed. Y. Terzian, New York, Benjamin, p. 571, 1968.

Czyzak, S.J., Krueger, T.K., Martins, P. de A.P., Saraph, H.E., Seaton, M.J., M.N.R.A.S. 148, 361, 1970.

Danks, A.C., Meaburn, J., Astrophys.Space Sc. 11, 398, 1971.

Day, G.A., Caswell, J.L., Cooke, D.J., Austr.J.Phys. Astrophys.Suppl. 25, 1972.

Dyson, J.E., Astrophys.Space Sc. 1, 388, 1968a.

Dyson, J.E., Astrophys.Space Sc. 2, 461, 1968b.

Eissner, W., Martins, P. de A.P., Nussbaumer, H., Saraph, H.E., Seaton, M.J., M.N.R.A.S. 146, 63, 1969.

Felli, M., Churchwell, E., Astron.Astrophys.Suppl. 5, 369, 1972.

Fischel, D., Feibelman, W.A., Ap.J. 180, 801, 1973.

Georgelin, Y.P., Georgelin, Y.M., Astron.Astrophys. 6, 349, 1970.

Goss, W.M., Shaver, P.A., Austr.J.Phys., Astrophys.Suppl. 14, 1970.

Habing, H.J., Israel, F.P., de Jong, T., Astron.Astrophys. 17, 329, 1972.

Hills, R.E., Janssen, M.A., Thornton, D.D., Welch, W.J., Ap.J.Lett. 175, L59, 1973.

Hoffman, W.F., Frederick, C.L., Emery, R.J., Ap.J. 170, L89, 1971.

Israel, F.P., Habing, H.J., de Jong, T., Astron.Astrophys. in press, 1973.

Isreal, F.P., Van der Kruit, P., submitted to Astron. Astrophys., 1973.

Kaler, J.B., Aller, L.H., Bowen, I.S., Ap.J. 141, 912, 1965.

Kleinmann, D.E., Ap.J.Lett. 13, 149, 1973.

Kleinmann, D.E., Low, F.J., Ap.J. 149, L1, 1967.

Kutner, M., Thaddeus, P., Ap.J.Lett. 168, L67, 1971.

Lemke, D., Low, F.J., Ap.J.Lett. 177, L53, 1972.

Mathews, W.G., O'Dell, C.R., Ann.Rev.Astron.Astrophys. 7, 67, 1969.

Mathis, J.S., Ap.J. 159, 263, 1970.

Mathis, J.S., Ap.J. 167, 261, 1971.

Morgan, L.A., M.N.R.A.S. 153, 393, 1971.

Münch, G., Persson, S.E., Ap.J. 165, 241, 1971.

Ney, E.P., Allen, D.A., Ap.J. 155, L193, 1969.

Ney, E.P., Strecker, D.W., Gehrz, R.D., Ap.J. 180, 809, 1973.

O'Dell, C.R., Hubbard, W.B., Ap.J. 142, 591, 1965.

Olnon, F.M., Habing, H.J., in preparation, 1973.

Osterbrock, D.E., Flather, E., Ap.J. 129, 26, 1959.

Peimbert, M., Goldsmith, D.W., Astron.Astrophys. 19, 398, 1972.

Pengelly, R.M., M.N.R.A.S. 127, 145, 1964.

Petrosian, V., Ap.J. 159, 833, 1970.

Petrosian, V., Silk, J., Field, G.B., Ap.J. 177, L69, 1972.

Pottasch, S.R., invited lecture, IAU regional Symposium, Athens, 1972.

Raimond, E., Eliasson, B., Ap.J. 155, 817, 1969.

Reifenstein, E.C., Wilson, T.L., Burke, B.F., Mezger, P.G., Altenhoff, W.J., Astron.Astrophys. 4, 357, 1970.

Riegel, K.W., Ap.J. 148, 87, 1967.

Rodgers, A.W., Campbell, C.T., Whiteoak, J.B., M.N.R.A.S. 121, 103, 1960.

Rubin, R.H., Turner, B.E., Ap.J. 165, 471, 1971.

Ryle, M., Nature 239, 435, 1972.

Schraml, J., Mezger, P.G., Ap.J. 156, 269, 1969.

Scoville, N.Z., Solomon, P.M., Ap.J. 180, 31, 1973.

Seaton, M.J., Progress Rep.Phys. 23, 313, 1960.

Sharpless, S., Ap.J.Suppl. 4, 257, 1959.

Solomon, P.M., Physics Today, March issue, 32, 1973.

Strömgren, B.E., Ap.J. 89, 526, 1939.

Turner, B.E., Scientific American, March issue, 51, 1973.

Wilson, O.C., Münch, G., Flather, E.M., Coffeen, M.F., Ap.J.Suppl. 4, 199, no.40, 1959.

Wilson, T.L., Mezger, P.G., Gardner, F.F., Milne, D.K., Astron.Astrophys. 6, 364, 1970.

Wink, J.E., Webster, W.J., Altenhoff, W.J., Astron.Astrophys. 22, 251, 1973.

Winnberg, A., Habing, H.J., Goss, W.M., Nature Phys.Sc. 243, 78, 1973.

Wurm, K., Rosino, L., Monochromatic Atlas of the Orion Nebula, Osservatorio Astrofisico, Asiago, 1959.

Wurm, K., Rosino, L., Monochromatic Atlas of the Orion Nebula, Suppl., Osservatorio Astrofisico, Asiago, 1965.

Wynn-Williams, C.G., M.N.R.A.S. 151, 397, 1971.

Wynn-Williams, C.G., Becklin, E.E., Neugebauer, G., M.N.R.A.S. 160, 1, 1972.

Wynn-Williams, C.G., Becklin, E.E., Neugebauer, G., preprint, 1973.

PHYSICS OF INTERSTELLAR MEDIUM

S.R. Pottasch

University of Groningen
Groningen, The Netherlands

I. GOALS IN STUDYING THE "NEUTRAL" INTERSTELLAR MEDIUM

This material plays a part in the evolution of our galaxy. It can absorb energy which otherwise might escape; it plays a role in the formation of stars; it has, at least locally, gravitational effects. Further it interacts with stars and stellar radiation.

In order to study these effects we must know how much material there is, in what form it is, how it radiates and absorbs energy. In particular we wish to know the kinetic temperature, T_e, the neutral hydrogen density, n_H, the electron density, n_e, the velocity fields, the dust concentration, and the abundance, all as function of the position in the galaxy. Knowledge of these parameters allows a discussion of the ionization and energy balance, which leads to a knowledge of the source of energy input.

Although in the past ten years large strides have been made in our knowledge of this subject, we are still at present unable to specify any of these parameters very precisely, nor do we know the primary source of energy input into the interstellar medium. It is the purpose here to summarize the present state of knowledge and the problems which are present. We will do this in the following way. First we will list the observations which are available of the interstellar medium and discuss the physics involved in the interpretation of each of these measurements. Secondly we will discuss the physical processes which are associated with "cloud" structure, thermal instability, and energy loss, because these are necessary for the further discussion. Thirdly, we will use these observations, together with their physical interpretation, to try to derive a consistent picture of the interstellar medium, or at least to see how far one can get in this direction before several interpretations are possible.

II. SUMMARY OF IMPORTANT OBSERVATIONS

A. List of observations which have been made of the interstellar medium

1. Interstellar absorption lines, in various parts of the spectrum, absorbed by atoms, ions and molecules.
2. 21-cm hydrogen lines, both in emission and absorption.
3. Pulsar dispersion measure.
4. "Weak" Hα emission.
5. "Weak" radio hydrogen recombination lines.
6. Radio continuum, free-free emission and absorption.
7. Emission lines of other elements, and molecules.

B. Detailed discussion of these observations, including the physics involved in the interpretation

1. Interstellar absorption lines. These lines were first observed by Hartmann (1907) in the spectra of δ Orionis. They could be identified as interstellar rather than stellar because this star is a spectroscopic double star, and the interstellar lines do not partake in the Doppler motion of the stellar spectral lines. An interpretation as circumstellar is also possible, and it was not until the work of Plaskett and Pearce (1933) that it was definitely established that the lines were interstellar, principally by establishing that lines of the same, non-stellar velocity were present, in roughly the same strength, in stars several degrees apart on the sky. Further, there is a correlation between the strength of the interstellar line and the distance of the star.

Most of our present information comes from the interstellar lines in the visual part of the spectrum, in particular Ca^{+} and Na^{0}. In the near future the ultraviolet results, especially of the Princeton group, will become available, which will greatly extend our knowledge.

The absorption lines which are observed mostly originate from the ground state, since in the very low density conditions in the interstellar medium almost all the population of an atom or an ion is in the ground state. A few absorption lines are observed from levels above the ground state with long lifetimes, and these will be very important for determining the conditions in the gas.

The analysis used in going from a measured equivalent width for a given component to a line-of-sight or column density is the curve-of-growth analysis, discussed first by Unsöld and Minnaert and applied to the interstellar lines by Strömgren (1948). The principle is as follows: when the number of absorbing atoms is small, the line is weak and every atom is able to cause an absorption. The equivalent width $W\lambda$ is then directly proportional to the number of absorbing atoms:

$$N = \frac{W\lambda \; 10^{-11} \, mc^2}{\pi e^2 \lambda^2 f} = 11.3 \frac{W\lambda}{\lambda^2 f} \, cm^{-2}$$

where W_λ is in milli-Angstroms and the wavelength λ is in cm.

Every atom can cause an absorption until all the light in the line center has been absorbed. The line is then referred to as being saturated. When this point occurs it depends on how large the motions of the particles are, i.e. how large the Doppler width of the line is[+]. After saturation begins the strength of the line increases only very slowly with the number of absorbing atoms, because there are only very few atoms in the Maxwellian tail which can absorb far enough from the line center where there is still radiation to be absorbed. However, when a large enough number of atoms is present, the "damping" wings of the line become important and the increase of the W_λ becomes more rapid with increasing N. In the interstellar medium this "damping" does not occur, as in stellar atmospheres, by the interaction of the absorbing particle with the electromagnetic field of a neighboring particle (pressure broadening) because the density is so low that there are very few neighboring particles. The "damping" occurs because of the so-called "natural broadening" or "radiation damping" which physically can be visualized as an inherent uncertainty in the position of the energy level as a consequence of the

[+]It is assumed in this analysis that the absorbing gas is uniform, and has a Maxwellian distribution of particle velocities.

Heisenberg uncertainty principle. In the laboratory, however, pressure broadening is much more important. The relation between W_λ and N for natural damping is (e.g. Unsöld, 1955):

$$W = \frac{2\pi e^2}{mc^2} \lambda \left\{ 2N \frac{f_{Lu}}{g_u} \sum_l g_l f_{lu} \right\}^{\frac{1}{2}}$$

where the transition occurs from the lower level L to the upper level u. The upper level is broadened by spontaneous decays to the lower level l. The statistical weight is denoted by g.

Fig. 1 gives an example of a curve-of-growth. As can be seen, for unsaturated lines and for lines on the damping portion of the curve, a measured equivalent width leads directly to a column density and is independent of

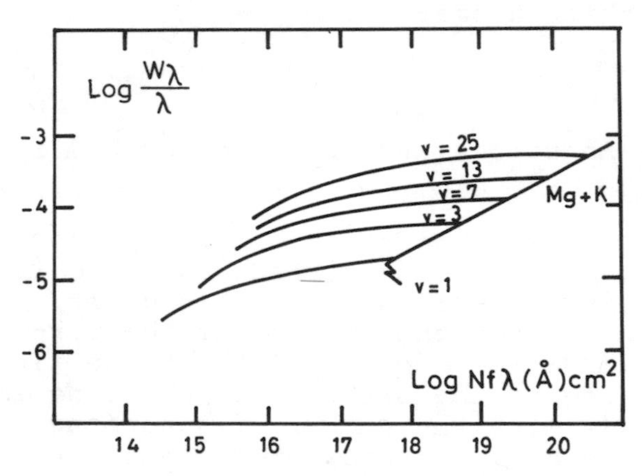

Fig. 1: Curve-of-growth for spectral lines of atoms at various mean velocities v (km/sec).

the velocity field in the gas. For the intervening region a single line measurement W_λ does not unambiguously give N. We must measure more than one line from a given level. There are such doublets and triplets available (they must have different f values) and then both N and the velocity v can be found. A word of caution is in order: the ratio of the equivalent widths for these two lines must be accurately measured, otherwise it could lead to large errors in the column density N.

Further we should stress that this analysis applies to a uniform gas with a specified Maxwellian velocity v. If there are several "components" we can only apply this analysis if these "components" can be separated, which is not always the case.

In this way we determine the column density of a given ion. This may be essentially the total density of the entire species, or much of the species may be in a different state of ionization. We shall return to this under C, the discussion of interstellar radiation field.

2. 21-cm hydrogen, emission and absorption. This transition was first studied by van de Hulst almost thirty years ago. It is due to a change in the relative orientation of the spin of the electron and of the proton, from the case in which the spin axes are parallel to that in which they are anti-parallel (the latter case has the lowest energy). The spontaneous radiative transition rate is very small, $2.85 \times 10^{-15} s^{-1}$. This is important, since it is substantially smaller than the electron collision rate, which is about $10^{-9} N_e$. This means that the levels will be populated as in thermodynamic equilibrium, unless the absorption of Lyman α radiation can preferentially depopulate one of these levels. This has been investigated and it appears (van Bueren and Oort, 1968) that only under very special conditions will this make a difference. Normally the thermodynamic equilibrium populations will obtain.

If this is true, the population of the emitting (upper) level will be essentially independent of the temperature, as long as the temperature is higher than $1^{\circ}K$. This means that the intensity of the 21-cm emission line, in the optically thin case, will give only information about the neutral hydrogen density and its distribution over the line-of-sight. For the optically thin case we can express the column density of neutral hydrogen as

$$N_H = 1.82 \times 10^{18} \int T_B (v)\, dv$$

where T_B is the observed brightness temperature in $^{\circ}K$ and v is the velocity in km s^{-1}. No information about the temperature is available in this case.

The amount of information obtained about the neutral hydrogen distribution in the past twenty years is impressive. It partakes in the general galactic rotation. It is concentrated in spiral arms where its volume density $n_H \sim 2$ cm^{-3}. It may be present between spiral arms. The neutral hydrogen is confined to the galactic plane, having an effective thickness of about 220 pc, but some neutral hydrogen appears to be still present at quite high heights above the galactic plane (the density $n_H \sim 0.01\ cm^{-3}$ may be found at between 1 and 2 kpc from the galactic plane (Kepner, 1970)). The total mass of neutral hydrogen is about $5x10^9\ M_\odot$, or about 5% of the total mass in the galaxy.

Information concerning the temperature of the neutral hydrogen can only be obtained when the gas has measurable optical depth. Initial estimates of the temperature were based on the following line of reasoning: If we can assume the temperature is constant in some direction, it may be optically thick. The highest brightness temperature measured was 125°K (later 135°K), so that this must be the temperature of the gas. The only other means of obtaining the temperature from an emission line is to measure the width of the line profile. Since gas motions also broaden the line, this method is only useful in giving a lower limit to the temperature. Usually this lower limit is not significant, but in a few cases it is. For example, a few clouds are known to have temperatures less than 30°K, from this method.

The principal means of measuring the temperature is to measure the 21-cm line in absorption against a source of continuum emission. The optical depth τ_V, which is given by

$$\tau_V = 5.49 \times 10^{-14} \frac{N_H \phi_V}{T}$$

where $\phi_V\, dv$ is the fraction of atoms whose radial velocities lie within the range dv, can thus be measured. Since N_H can be measured from the emission in the line, T can be determined. One of the worst difficulties with this method is that one cannot measure emission and absorption at the same point. One can average the emission measured at neighboring positions in the sky, but this is not always satisfactory. Measurements with an interferometer (Radhakrishnan et al., 1972) can perform this averaging in a more satisfactory way.

It is clear that if other things are equal, the optical depth will be higher for lower values of the temperature. Thus, this method will have a systematic bias in favor of measuring the low temperatures. The resulting temperatures will be discussed in the last section.

3. Pulsar dispersion measures. The group velocity v_g of a radio wave in ionized hydrogen is given by

$$v_g = c\left(1 - \frac{v_p^2}{2v^2}\right)$$

when the plasma frequency v_p is much less than the wave frequency v . The plasma frequency depends on the electron density n_e

$$v_p^2 = \frac{n_e e^2}{\pi m}$$

It follows that the travel time of the photon depends on the frequency and n_e. With a constant source of energy we cannot measure the difference in travel time at different frequencies. But if we have a source which flashes on and off, such as a pulsar, we can accurately measure the difference in travel time. Since the travel time depends on the velocity and the distance, its measurement determines the quantity $\int n_e dl$ along the line-of-sight. This is called the dispersion measure and is usually measured in units of pc cm^{-3}.

The question immediately arises as to whether this dispersion occurs in the immediate vicinity of the pulsar, or whether it occurs in the general interstellar medium. This can be decided by looking to see if there is a dependence of the dispersion measure on galactic coordinates. If we divide the sixty known dispersion measures in five groups based on variation in the galactic latitude b, we obtain the following result:

Galactic latitude	$\int n_e dl$
$\lvert b \rvert < 2^\circ$	142
$2^\circ < \lvert b \rvert < 5^\circ$	60
$5^\circ < \lvert b \rvert < 10^\circ$	59
$10^\circ < \lvert b \rvert < 30^\circ$	37
$30^\circ < \lvert b \rvert < 90^\circ$	13

This is convincing evidence that at least the major part of the dispersion measure is formed in the general interstellar medium.

4. "Weak" Hα emission. In addition to the hydrogen Balmer line emission from the HII regions, there may be a general low intensity emission from the general interstellar medium. This has been observed by Johnson (1972) who found "weak" H α emission along the galactic plane. This has been investigated in more detail by Reynolds, Roesler, and Scherb (1973), especially in Hα emission.

These authors were able to isolate regions of the sky six minutes in diameter, and measure the profile and radial velocity of the observed emission. Some of their results are shown in Fig. 2. The contours shown on this figure are the intensities of the neutral hydrogen in

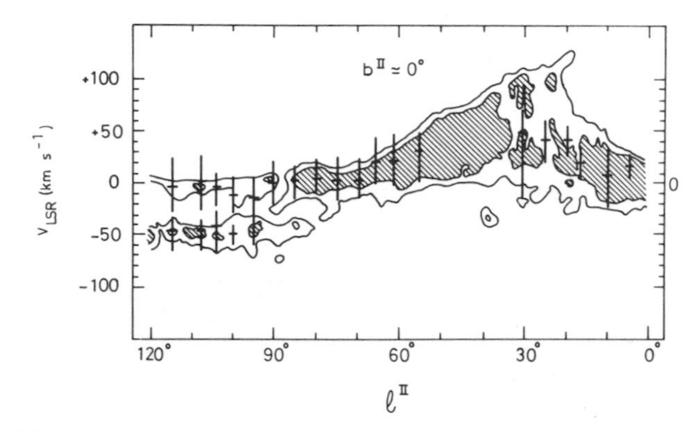

Fig. 2: Radial velocity from Hα emission in the galactic plane (Reynolds et al., 1973). Contours are intensities of neutral hydrogen as a function of velocity.

the galactic plane as a function of the radial velocity. The lines indicate measurements of Hα . It is striking that the Hα coincides almost exactly with the neutral hydrogen, i.e. it is possible to follow the spiral structure, at least in the three nearest spiral arms, in the Hα radiation. No data is plotted between 30° and 55° longitude because the intensity of the emission was too low to accurately determine the peak position. This region of low intensity corresponds to the interarm region between the Sagittarius and Orion spiral arms. Normal HII regions have been excluded from this plot. The intensity of this radiation is known in an approximate way.

This radiation is probably formed by radiative recombination of protons and electrons, followed by cascade to lower levels. This process is well studied (e.g. Seaton, 1960; Burgess, 1958; Brocklehurst, 1970). The intensity is given approximately as

$$I(H\alpha) = 3.2 \times 10^{-21} \frac{n_e n_p}{T_e} \text{ erg cm}^{-3} \text{ s}^{-1}$$

in the temperature range from 5000° to 20000°. Below 5000° it increases less strongly; approximate calculations by Grewing indicate that the ratio of the intensities at lower temperatures are as follows:

T	Relative $H\alpha$ radiative recombination intensity
5000°	1
1000°	2.9
100°	10.6
40°	17.1

There is at present no way of estimating how much of the $H\alpha$ emission is being absorbed by the interstellar dust.

5. "Weak" radio hydrogen recombination lines. The same recombination between protons and electrons which produced the $H\alpha$ emission, also produces a large number of other recombination lines. Those originating in transitions between the higher levels have little energy and occur in the radio region of the spectrum. The temperature dependence of this radiation is slightly different than for $H\alpha$ emission:

$$T_L \Delta\nu_L = 1.93\times10^3 \, T_e^{-3/2} \, e^{X_n} \int n_p n_e \, dl$$

where the intensity is expressed in terms of a measured brightness temperature in the line, T_L, and the measured half power half width $\Delta\nu_L$ (kHz). The $X_n = \frac{157800}{n^2 T_e}$, where n is the quantum number of the radiating level, and l is the line-of-sight in parsecs.

Measurements of this radiation have been made by Gottesmann and Gordon (1970) (H158α), Jackson and Kerr (1971) (H109α), and Davies, Matthews and Pedlar (1972) (H166α). A summary of the observed radiation is shown in Fig. 3. It is similar to Fig. 2 in that the velocity extent of the observed radiation is superimposed on the neutral hydrogen contours. It can be seen that the radiation originates in a very similar region of space. In making the measurements, an attempt has been made to avoid those directions where known HII regions are present, so that, as in the case of $H\alpha$ emission, we are measuring the background radiation.

6. Radio continuum. An electron passing through the electric field of a neighboring proton may be slightly decelerated causing the emission of so-called free-free radiation. Conversely, radiation may be absorbed by the same mechanism. This effect is especially of im-

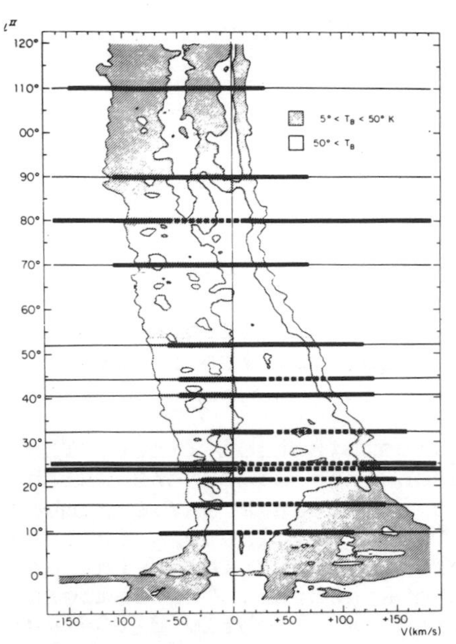

Fig. 3: Radial velocity from "weak" radio recombination lines in the galactic plane, with neutral hydrogen contours.

portance in the radio region of the spectrum.

The free-free emission background is observationally important, but very difficult to be observed. This is because there is, in addition, at the same radio frequencies, a non-thermal background, which is very difficult to separate from the free-free background. Attempts have been made to effect this separation on the basis of the fact that the frequency dependence of these two sources of emission is different. Unfortunately the frequency dependence of the non-thermal source is not precisely predictable, so that present opinion varies as what percentage of the radio frequency background is due to free-free emission.

The absorption due to free-free transition has been observed. At radio frequencies, the optical depth may be written

$$\tau = 8.24 \times 10^{-2}\ T_e^{-1.35}\ \nu^{-2.1} \int n_p n_e\, dl$$

where the frequency ν is expressed in units of GHz and l is in parsecs. Thus, the absorption increases strongly toward longer wavelengths. Several authors (e.g. Ellis and Hamilton, 1966; Bridle, 1969) have measured absorption occurring in the interstellar medium in the frequency range near 10 MHz and below. Bridle reports an absorption coefficient $\chi = 0.34$ Kpc^{-1} in the galactic plane and an optical depth $\tau = 0.1$ toward the galactic pole, both at a frequency of 10 MHz. Ellis and Hamilton give a similar value at this frequency: $\tau = 7.6 \times 10^{-1}$ at $|b| = 5^{\circ}$ and $\tau = 3.7 \times 10^{-2}$ at $|b| = 60^{\circ}$. Goss (1972) and Gordon (1972) have suggested that at least in the direction of 3C 391 ($l = 31^{\circ}.9$, $b = 0^{\circ}.0$) the absorption may be higher: a value of (10 MHz) = 60 is suggested for this object, which has a distance of the order of 10 Kpc. This has been confirmed by Dulk and Slee (1972), who find comparable absorption in the direction of fifteen supernova remnants within 40° of the galactic center.

7. Emission lines of elements other than hydrogen. (a) Atomic ionic lines. Radiation from these lines should be the principal source of cooling in the interstellar medium, at values of $T_e < 10^4$ °K. The cooling (and production of these emission lines) occurs because a particle collides with the atom or ion, raising it to an excited state from which it radiatively decays to a lower level, emitting the line in question. Thermal electrons are one of the important possibilities for the colliding particle. In that case, when the temperature of the gas is low, it will only be possible to excite energy levels that are close to the ground level, since most of the electrons do not have enough energy to excite the higher levels. For example, at temperatures above 40°K, the $^2P_{3/2}$ level of CII may be excited by electron collision, and thereafter radiate a line at 156μ . Other lines of importance, and the temperatures at which they will become important ($\Delta E/kT < 2$) are:

	$\frac{\lambda}{\mu}$	$\frac{T}{°K}$
C^0	609.8	12
N^+	203.6	35
N^+	121.6	95
O^0	63.1	110
Si^+	34.8	200
H_2	28.2	250
Fe^+	26.0	280
Fe^+	22.9	310

	λ / μ	T / °K
Ne^+	12.8	560
Fe^+	5.3	1300
N^+	0.6584	11000
O^+	0.3727	20000

The strength of the lines (and the importance of the cooling) will depend on the collision cross-sections, the temperature, and the abundance of the ion in question. Unfortunately, the lines which are expected for the low temperature gas lie in the infrared part of the spectrum and have not yet been observed. It is clear that their eventual observation will be an important addition to our knowledge.

These levels may be also excited in a thermal gas by collisions with neutral hydrogen atoms. This process is only about 1% as efficient as electron collisions, or less, but it may be of importance in regions where the ionization is very low. Collisions by neutral hydrogen atoms have a different temperature dependence than electron collisions, in the sense that the maximum rates occur at higher temperatures for the neutral atom collisions.

These are probably the most important ways of populating the excited levels, and for cooling they are probably the only processes that need to be considered. It may be, however, that other processes are important for excitation in specialized conditions. We list some possibilities:

(1) absorption of radiation, followed by cascade;
(2) direct excitation by non-thermal particles (e.g. cosmic rays);
(3) charge exchange reactions, leaving one of the products in an excited state;
(4) excitations by secondary electrons formed as a result of other non-thermal particle collisions.

Some of these processes have been discussed in the literature (e.g. Bergeron and Souffrin, 1971), but not exhaustively.

The only measurements which have been made of emission lines up to the present are those of Reynolds et al., 1973, who report that the (NII) λ 6584 line is often observed at the same position as the "weak" Hα, with the same radial velocity; presumably it originates from the same gas. The intensity of (NII) is less than 30% of the Hα intensity. In many extragalactic objects this same extended "weak" Hα is seen (Monnet, 1971). Sometimes (M33 and M101) the (NII) lines are also weakly seen. In unpublished work other lines (OII) and (SII) are seen as well, both in the arms and between them. This may also be true in our galaxy.

Concerning molecular lines we have already mentioned the possibility of observing the line at 28.2μ. In regions with temperatures above several hundred degrees, it should also be possible to measure the lines at 12.3μ and 8.03μ, while at higher temperatures the lines at 6.10μ and 4.40μ may be observable.

Most of the molecules observed at present in the radio region refer to regions in the interstellar gas where very special conditions hold. Probably only a few have been observed in "normal" regions, and these only in dust clouds. Usually CO, H_2CN, and OH are observed. We shall not discuss dust clouds in detail here, other than to say that the molecules probably are at a low temperature ($\sim 10^oK$) in these clouds.

C. Ionization equilibrium and the interstellar radiation field

Ionization can occur in the interstellar medium by any of the following processes:

(1) absorption of radiation of sufficient energy;
(2) collision by thermal electrons;
(3) collision by non-thermal particles (cosmic rays);
(4) charge exchange processes.

Collision by thermal electrons, process 2, will begin to be important, for the elements which interest us, at temperatures above 8000°K. The cross-sections for this process are usually sufficiently well-known (see Bely and Van Regemorter, 1972).

The absorption of radiation is undoubtedly an important process. Here we must make a distinction between the radiation field on the longwave side of the Lyman series limit (λ = 912 Å) and the shortwave side. This is because the neutral hydrogen column density, N_H, can reach values of 10^{20}-10^{21} cm^{-2} in and near the galactic plane. Since the absorption coefficient at the Lyman limit is 6.3 x 10^{-18} cm^2 the optical depth can be $\sim 10^3$ and only the very nearby radiation field will be important, immediately shortward of the Lyman limit. Since the absorption coefficient varies as the cube of the wavelength, below 100 Å this situation is again completely changed, and the X-ray radiation may play an important role. This is, however, not at all certain, because the X-ray spectrum is not well measured in this wavelength range. The X-ray flux is reasonably well-known shortward of 10-20 Å, but here it is not important in influencing the ionization equilibrium.

The radiation longward of 912 Å is clearly important for ionization of atoms and ions with an ionization potential less than 13.6 e.v. The radiation field is probably due to the sum of the radiation fields of the individual stars, modified by the intervening extinction by dust. Because we are interested mainly in the radiation field from between 912 Å and 2000 Å, the hot stars are primarily responsible. Thus, we must first determine the space density of the hot stars. Secondly, we compute or measure the radiation of each of these stars as a function of the spectral type. We then combine these two pieces of information to determine the average radiation field in the solar neighborhood. We must modify this field to take into account absorption by the intervening dust.

This calculation was first made by Dunham (1939). It was improved by Habing (1968) who used more modern star distributions, as well as improved models for the radiation fields of individual stars. It has now been further improved by Witt and Johnson (1973), who have used the OAO measurements of hot stars to revise the radiation field. The resultant radiation field is shown in Fig. 4.

The possibility that cosmic rays contribute to the ionization has been discussed in the literature for some time. It was clear more than twenty years ago that cosmic rays of an energy higher than 1 GeV are not observed in sufficient number to importantly contribute either to the ionization or to the heating of the interstellar medium. The limit of 1 GeV is an observational one: lower energy particles do not penetrate the atmosphere. In 1960, Hayakawa et al., noting a possible discrepancy between the predicted temperature of the interstellar medium and the "observed" temperature, renewed the suggestion that lower energy cosmic rays are an important source of ionization and heating. In recent years, this has received support from theoretical work which shall be discussed shortly.

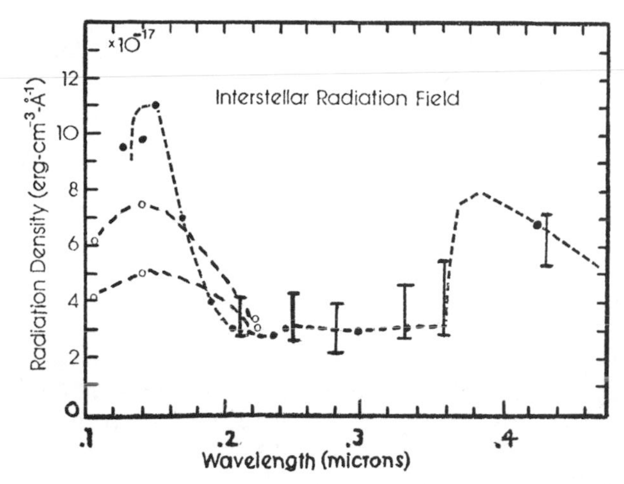

Fig. 4: Interstellar radiation field (Witt and Johnson, (1973).

Also observations of cosmic rays from satellites have eliminated the effects of the earth's atmosphere. This has allowed observations to be made almost to energies of 1 MeV, but one is not measuring the interstellar cosmic rays at these low energies. This is because the interplanetary medium (and especially the "turbulent" magnetic fields in the interplanetary medium) prevent the low energy particles from reaching the earth. While theoretical work has been performed to determine a correction for the interplanetary medium, the present status of the problem is that it is too difficult to give a correction, because the interplanetary medium acts not only to keep out low energy cosmic rays, it also changes the energy of individual particles. Thus, we must use indirect arguments to determine the cosmic ray intensity.

The final process, charge exchange, may play a role in certain cases. For example, the process

$$H^+ + O^o \rightleftarrows H^o + O^+$$

appears to have a large reaction rate (Field and Steigman, 1971). This may be because O^o and H^o have almost the same ionization potential. The general theory is at present so poorly understood that calculated reaction rates for atoms of substantially different ionization potentials have uncertainties of the order of 10^3. We therefore must be content at present to realize that this possibility exists.

We have now discussed the ionization processes. The recombination processes which are important in the interstellar medium are two-fold:

(1) direct radiative recombinations in which an electron and an ion physically combine, emitting a photon in the process. For hydrogen-like ions the recombination rate is well-known. Total recombination rates are usually reliably known because much of this total is recombination to high quantum levels which are hydrogenic.

(2) dielectronic recombinations in which the recombination can be thought to take place in two steps. First the electron and ion for an atom in an excited state corresponding to two electrons being excited and having the same energy as the free particles. This transition does not involve the emission of radiation. From this highly excited state it can return to the original configuration, or it can cascade to a lower level, emitting radiation in the process, and becoming bound. The rates for this reaction have been given by Burgess (1965). It will be important for the interstellar medium at temperatures of the order of 10^4 °K.

III. "CLOUD" STRUCTURE, THERMAL INSTABILITY AND ENERGY BALANCE

A. "Cloud" structure

One of the first things which is observed in a photograph of the sky is the irregular distribution of the dark matter. This is the origin of the idea that the interstellar medium is not homogeneously distributed, but is in the form of irregular "cloud" structures. This structure of the dark matter also provided one of the first estimates of the average distribution of the "clouds": the "clouds" occupy 5% of the volume of space. This value is still used today. But there is other evidence for "clouds". Chief among this evidence are the observations of the interstellar absorption lines. First of all the lines, when they are not clearly saturated, are often quite narrow, usually 2-4 km s^{-1} as a half-width (Hobbs, 1969; Marschall and Hobbs, 1972). Secondly, if the distance to the star is of the order of 500 pc, two or three or sometimes more of the narrow components are seen. The velocities may be spread over more than 30 km s^{-1}, clearly indicating that these are separate gas masses.

In 21-cm neutral hydrogen these structures are not usually seen, at least in the galactic plane. The reason for this is not entirely clear. It may have something to do with the fact that observations are made through a much longer line-of-sight in the 21-cm emission line. At latitudes $|b| > 15^o$ some cloud structures are seen in neutral hydrogen. They have sizes of 10°-20° on the sky, but their distances are not well-known.

The size of these masses is difficult to determine, and statistical methods give an unsatisfactory answer, in the sense that a large range of sizes may exist in reality and an average value, weighted according to an unknown function, has a very uncertain meaning. Nevertheless, estimates are available in the literature, the most quoted being that of Spitzer (1968), who estimates the size of a "standard cloud" as 7 pc and a "large cloud" as 30 pc.

Is there anything between these gas masses or "clouds"? This is a question which has been posed as far back as twenty-five years ago (e.g. Strömgren, 1948) and a satisfactory answer is not yet available. One began by asking how these "clouds" can remain as entities over a long period of time. It is possible if they are in pressure equilibrium with a surrounding "intercloud" medium. This also gives an estimate of the properties of this "intercloud" medium. Strömgren, for example, assumed a temperature of 10^2 ^oK for the "clouds" and 10^4 ^oK for the "intercloud" medium, for reasons which will be discussed shortly which makes the "intercloud" medium 1% as dense as the "clouds".

But there is no compelling reason to assume that the "clouds" are in equilibrium with the surroundings. There could just as easily be a dynamic equilibrium in which these "clouds" are constantly changing under the influence of their internal pressure, the ultraviolet radiation incident on them, their mutual interaction (collision), etc.

Perhaps the most important piece of evidence which may have a bearing on this problem is the comparison of the profiles of 21-cm emission and absorption in the same region of space. Fig. 5 is an example of such a comparison. As can be seen the absorption line is always narrower than the emission line. This was first noted by Clark (1965) and is confirmed by later, and better, observations. Radhakrishnan et al. (1972) have shown that

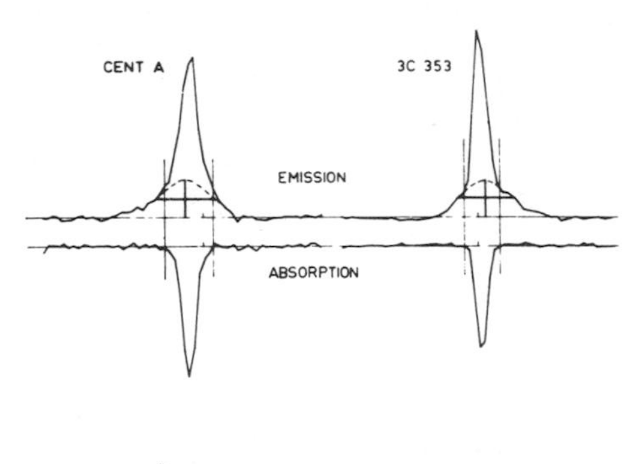

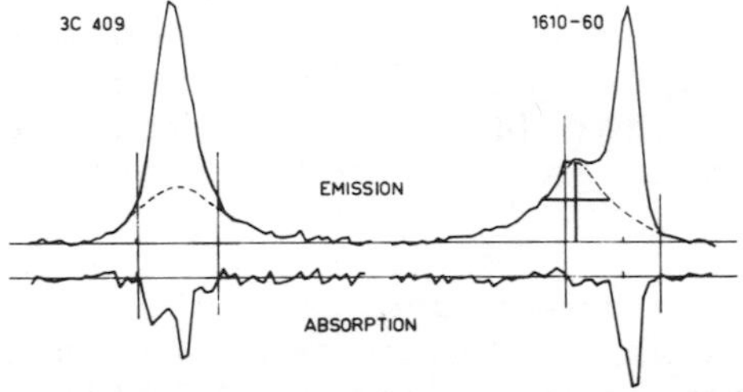

Fig. 5: Comparison of line profiles in emission and absorption around 21 cm.

if a gaussian component analysis is made, there is a wide component that appears in emission, but not in absorption, and that the strength of this component is inversely proportional to the sin $|b|$. This can be thought of as a gas of higher temperature. Only lower limits on the temperature can be placed, and these vary from $150^{o}K$ to $750^{o}K$, with a mean about $400^{o}K$. The width of this gaussian component varies between 19 and 50 km s^{-1}, which allows the possibility that the actual temperature may be higher than 10^4 ^{o}K.

I think that one may deduce from these measurements that some fraction, perhaps a substantial fraction, is at a temperature greater than $400^{o}K$. Some authors have seen in these results a justification of a two-component model as described above, but this is probably premature, and it may be actually misleading.

B. Thermal instability in connection with cosmic ray heating

For each manner of ionizing and heating the gas we may write an equation of energy balance. These two equations relate n_H, n_e, and T_e, together with the flux of this energy. Since the cooling mechanism depends on ions such as C^+, Si^+, and Fe^+, it is also necessary to know the abundances of these ions, which will be discussed presently. A solution of these two equations (see e.g. Pikelner, 1967; Field, Goldsmith, and Habing, 1969; and the summary of Dalgarno and McCray, 1972) leads to relationships between these quantities. An example is shown in Fig. 6, where the upper part of the curve shows the pressure as a function of the total density. The

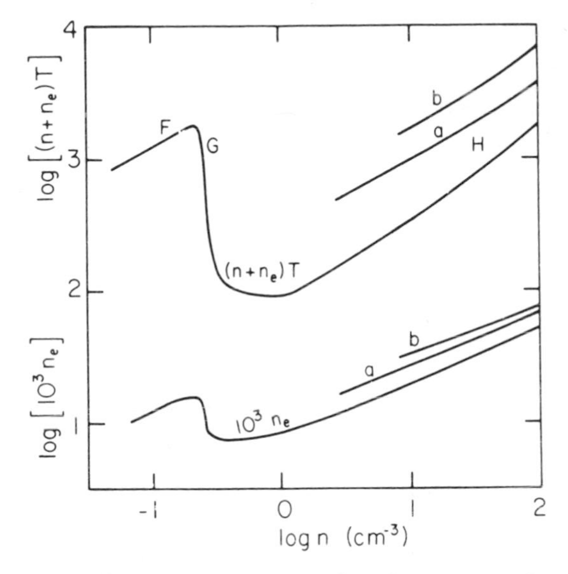

Fig. 6: Pressure and electron density as a function of total density of interstellar medium after various authors.

physical situation is stable only in that region in which the pressure increases when the density increases. In Fig. 7, the temperature is shown as a function of the density. It can be seen that in the stable region at low density the temperature is high, and the higher density stable region occurs at low temperatures. Field, Goldsmith, and Habing (1969) suggest identifying the high temperature region with an "intercloud medium" and the low temperature region with "clouds". It is not implausible but it is only a suggestion which must be confirmed by observation. Hjellming et al. (1969), following Field (1962) and Pikelner (1967), have suggested that the unstable region might be associated with an "intercloud medium". While this theoretical discussion should be used as a basis for thinking about what observations to make, they should not be allowed to bias the interpretation of observations.

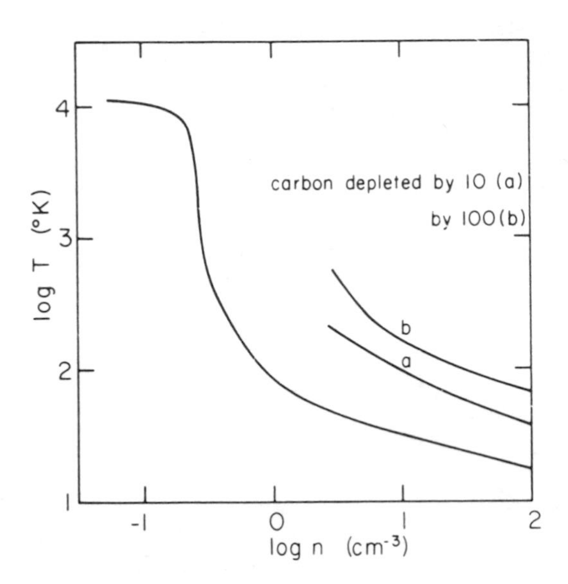

Fig. 7: Temperature as a function of total density of interstellar medium.

It is interesting to ask what physical reason there is for this instability. It is simply that when a gas is compressed the energy loss increases faster (n^2) than the energy input increases (n) and the temperature falls. There is a region where the temperature falls so quickly that compressing the gas actually causes the pressure to decrease so that the gas continues to contract.

It should be noticed in Fig. 7, where equilibrium temperatures are given, that for "clouds" of densities $n \sim 10$, temperatures of the order of 20°-30° are found. Only when the abundances are one or two orders of magnitude lower are temperatures as high as 100° predicted. When only radiative heating is present, a temperature of about 20^oK is predicted, independent of the density, and having the above dependence on the abundance. This makes the observed temperature distribution (see following section) difficult to understand, if the only energy input is radiation with $\lambda > 912$ Å.

IV. INTERPRETATION OF THE OBSERVATIONS IN TERMS OF THE STRUCTURE OF THE INTERSTELLAR MEDIUM

In this section we shall apply what has been discussed until now to try to obtain some idea of n_e, T_e, n_e/n_H, and the abundance. We shall divide the discussion into parts, discussing each of these in turn. Unfortunately the observations usually give some mixture of these quantities, integrated over the line-of-sight. The sections therefore overlap each other; further the inhomogeneity must always be borne in mind.

A. Electron density n_e

1. The electron density can be obtained from the ionization equilibrium as derived from the interstellar lines. Especially important at present is the calcium equilibrium, since both Ca^o and Ca^+ have been observed in about twenty components. In a given star, it appears that both of these components are observed at the same velocity so that they come from the same region of space. They originate in the "cloud" component, because these components are our definition of "clouds". As has previously been discussed

$$\frac{Ca^+}{Ca^o} = \frac{\Gamma}{\alpha(T_e)n_e}$$

Older observations have been summarized by Pottasch (1972a) and newer observations have been made by White (1973). They yield n_e only if T_e is known. For example, from the observations in the star HD 190066, we can obtain the following values of n_e:

T_e	n_e
100°	4×10^{-2} cm^{-3}
1000°	20×10^{-2} cm^{-3}
10000°	0.8×10^{-2} cm^{-3}

The reason for this behavior is that initially the radiative recombination rate decreases with increasing temperature, but at higher temperatures dielectronic recombinations become important and less electrons are needed to produce a given ionization ratio.

How to distinguish between these possibilities? There are different indications that the low temperature solution is the correct one. First of all, we can do this same analysis with Na. Unfortunately only Na^o is observed, but Na^+ can be estimated since almost all the Na is in the form of Na^+. Thus, if we know the amount of neutral hydrogen in the "cloud" and the abundance of Na, we have determined Na^+. The resultant values of n_e from this ionization equilibrium are very close to those determined from the Ca equilibrium for the lower temperatures. Above 3000° there is a large discrepancy, however, since di-electronic recombinations do not become important for Na^+ until much higher temperatures. We conclude that temperatures below 3000^oK obtain, and the values of n_e corresponding to the lower temperatures.

Secondly, as shall be seen presently, the 21-cm absorption temperatures also show that the stronger components have temperatures of the order of 100^oK.

Thirdly, the excitation temperatures found from H_2 measurements in the ultraviolet (OAO C), and which are probably kinetic temperatures, are almost always between 100° and 200^oK.

Only a limited number of CaI measurements have been made, because the line is very weak. It is observed only in the strongest interstellar components, and thus we have a strong systematic bias in favor of a certain result. Thus, we should not be surprised that all the "clouds" observed give very similar values of density and temperature. The values found are mostly in the range

$$5x10^{-2} \leqslant n_e \leqslant 2x10^{-1}$$

and the temperature probably lies in the range

$$100^o \leqslant T_e \leqslant 400^o.$$

Recently, measurements of the Mg^+ and Mg^o abundances have made it possible to use this ionization equilibrium as well. The results are completely consistent with the Ca and Na results, if one is very careful about using the curve-of-growth to obtain the line-of-sight densities. This is more difficult because of the strong saturation effects in the Mg^+ lines, and this has perhaps caused Boksenberg et al. (1972) to be in serious error. The results of De Boer et al. (1972) and Morton et al. (1973) are more reliable, and are consistent with the above values.

2. Pulsar dispersion measures. We observe $\int n_e dl$. If we know the distance to the pulsar, we can determine an average value $\bar{n}_e$. We now have a rough value of the distance to about eight pulsars, mainly by observing (or not observing) the 21-cm absorption line corresponding to a

spiral arm of which we know the approximate distance. For these few cases which lie close to the galactic plane we obtain a remarkably small range of value of $\bar{n}_e$, roughly

$$2\times10^{-2} < \bar{n}_e < 6\times10^{-2}$$

This value is similar to the electron density obtained from the Ca ionization equilibrium. It is difficult to draw any conclusion from this similarity because the value may not refer to the same gas. Especially, if the "clouds" in which the interstellar lines are formed are limited to a small fraction of the line-of-sight, it would mean that the value of $\bar{n}_e$ would refer to the rest of the material, which, by chance, has a similar electron density.

3. "Weak" Hα, free-free absorption measures, and hydrogen radio recombination lines. We group these measurements together because they are all proportional to $n_e n_p$ and a function of the temperature, as discussed earlier. They must all be formed in the same region of space because of the similar dependence on the electron density, and because, as discussed earlier, of a very similar variation of the emission along the galactic plane. Because of the somewhat different temperature dependence of these three processes, it is also possible to determine roughly the temperature of the emitting or absorbing gas.

Consider the approximate direction $l = 30^\circ$, $b = 0^\circ$, since all three observations have been made in this direction. This direction probably has a long path length through a spiral arm. We give in the following table the value of $n_e^2 \Delta l$ (cm^{-6} pc) derived from each of these measurements, assuming three possible temperatures.

T_e	$n_e^2 \Delta l$ from "weak" Hα emission	from free-free absorption τ(10 MHz)=60	from radio recombination line	Predicted free-free brightness temperature T_B at 6 cm
10^4	105.	12000.	3000.	1.3 °K
10^3	18.	520.	95.	0.13 °K
10^2	5.	23.	3.	0.013°K

The "weak" $H\alpha$ emission is not directly comparable to the other results because the interstellar absorption means that this emission is coming from a more localized region than the 10 kpc path length the other measurements refer to. Even if, however, the $H\alpha$ referred only to the nearest 2 to 3 kpc and a correction factor of three to five had to be applied to the column $H\alpha$ emission to make it comparable, we would still have to rule out a temperature $T_e = 10^4$ °K as not compatible with all of these observations. A value of between 10^2 and 10^3 is more likely, probably closer to 10^2. In that case $n_e^2 \Delta l \simeq 100$, which for a path length of 10 kpc, we find an average value of $n_e \simeq 10^{-1}$, which is surprisingly similar to the values found previously. This direction may be one of the above average absorption and emission.

We should stress that the conclusions drawn here are still controversial. Davies, Matthews, and Pedlar (1972), for example, claim to have separated the free-free background emission from the non-thermal background emission, and deduce a value of $T_e \simeq 6000$°K. In the last column of the above table the predicted brightness temperature due to free-free emission is given. Only if $T_e \sim 10^4$ °K can an important fraction of the observed continuum be due to free-free emission.

B. Electron temperature T_e

1. The only direct method of determining the value of T_e is from the simultaneous emission and absorption in the 21-cm hydrogen line. This has been done by Radhakrishnan et al. (1972) and Hughes et al. (1971) with the following distribution of T_e:

Temperature range	Number of "clouds"
20° - 40°	18
40° - 60°	26
60° - 80°	27
80° -100°	26
100° -120°	21
120° -150°	14
150° -200°	13
200° -300°	5

It is interesting to note the following points:

(1) The absorption becomes more difficult to be measured as the temperature increases, and only the largest masses are measured at the higher temperatures. Thus, the fact that no temperatures higher than 300°K have been measured may be observational selection.

(2) There appears to be no uniform temperatures for the gas which produces the absorption, but a spread of at least a factor of ten is observed.

(3) There are cases when the absence of absorption lines indicates temperatures above 300°-600°K. Further, the broader emission wings discussed earlier indicate the presence of such a higher temperature.

If we refer again to Fig. 7, we see that a temperature range from 20° to above 300° implies, on the basis of the theory of cosmic ray heating, a range of densities of more than two orders of magnitude.

2. The observation of the (NII) lines and other forbidden lines also offers the possibility of determining the temperature in the region in which it is formed, which is probably the same as in which the $H\alpha$ and the radio recombination lines are formed.

A first method of using this observation is to compare the profile of the line with that of $H\alpha$. The profile is determined by gas motions and by thermal broadening; since the effect of thermal broadening is different in H and N, it is in principle possible to separate these effects, at least if the gas is at a constant temperature.

Reynolds et al. (1973) claim to have done this in two cases and obtain a value of T=5000°K. On examining the observations on which this is based, it appears that the errors are so large, as to make the present determination extremely doubtful. The method is interesting and should be further exploited.

A second method is as follows: If we assume that the (NII) upper level is excited by collision with thermal electrons we can roughly estimate the temperature from the intensity of the line. However, there is reason to believe that processes, other than collisions, of thermal electrons are responsible for the excitation. The reason for this is that it appears from the measurements of the ultraviolet interstellar absorption lines (Morton et al.,

1973) that the ratio N^+/N^o is of the order of unity or somewhat greater. It is not yet clear what causes this high ionization of N (ionization potential 14.8 e.v.), but it is likely that whatever causes the ionization will also be effective in exciting this level. This must be studied carefully.

3. In summary it appears that the stronger interstellar lines are formed in regions of low temperature (100°-400^oK). This range of temperature is also consistent with the Hα measurements, the free-free absorption, and the radio recombination lines. A part of the 21-cm radiation comes from gas of this same temperature, but lower and higher temperatures are also present.

C. The state of ionization of hydrogen

This is an interesting point which is very difficult to be answered observationally. The difficulty is that the state of ionization may vary substantially from place to place in the interstellar medium and the averages obtained over the line-of-sight may not be meaningful. We shall present such averages in order to compare different kinds and see if any conclusion may be drawn.

(1) From the interstellar line ionization equilibrium we have shown that the electron density can be found. The density of neutral hydrogen can also be determined from these gas masses from the 21-cm measurements, if the line-of-sight is known or can be estimated. For six cases which are located in the spiral arms and which show very strong interstellar lines (see Pottasch, 1972b) a value $n_e/n_H \simeq 2x10^{-2}$ has been found. For the "clouds" studied by White a similar result has been found. But it may well be that some of the clouds have substantially different values, both lower and higher. For example, there are some interstellar line "components" which do not have corresponding 21-cm emission lines, which could mean that most of the hydrogen is ionized (it could also be interpreted as having most of the hydrogen in molecular form).

(2) Recently, Sancisi and Klomp (1972) have combined 24 pulsar dispersion measures with their 21-cm hydrogen measurements in the same direction. This comparison yields values of

$$\frac{\int n_e dl}{\int n_H dl} = \frac{N_e}{N_H}$$

These values will usually be lower than the actual situation because $\int n_e dl$ is only taken as far as the pulsar, and $\int n_H dl$ is taken over the entire line-of-sight, including the gas behind the pulsar. We should thus expect a variation with galactic latitude, since in the galactic plane a substantial amount of neutral hydrogen is expected to be behind the pulsar, while at higher latitudes most of the hydrogen will be in front of the pulsar, since the hydrogen is concentrated in the galactic plane. In fact, this is observed in the mean:

$$\text{for } |b| < 10^{\circ} \qquad \left\langle \frac{N_e}{N_H} \right\rangle \simeq 1 \text{ to } 2\text{x}10^{-2}$$

$$|b| > 10^{\circ} \qquad \left\langle \frac{N_e}{N_H} \right\rangle \simeq 1.5\text{x}10^{-1}$$

The value at the higher galactic latitudes is thus probably the value to be preferred, except in the case in which the state of ionization varied as a function of distance from the galactic plane.

(3) If it is true that the state of ionization as found by these two different methods is similar, it rules out, at least to a substantial extent, a model whereby one region was completely ionized and the other region had an ionization state $n_e/n_H \simeq 3\text{x}10^{-4}$. If we conclude that the average ionization of hydrogen is from 3 to 10%, we must conclude that a process other than ionization by the average stellar radiation field above 912 Å is important. This other process could be either of those suggested in an earlier section, cosmic ray ionization or X-ray ionization, although present soft X-ray intensity measurements make the latter process unlikely. The ionization state within a cloud allows us to determine ζ, the number of ionization of hydrogen per second. The value is 3 to $10\text{x}10^{-15}$ s^{-1}, which is much higher than can be produced by radiation, or by the observed cosmic ray or X-ray flux.

D. Neutral hydrogen densities

Measurements of 21-cm emission yield line-of-sight neutral hydrogen densities. If the gas were uniformly distributed with 100 pc on either side of the galactic plane, it would have an average density n_H = 0.4 cm^{-3}. It is certainly concentrated in spiral arms and most likely in somewhat smaller units, and average densities in these units may be 10 cm^{-3}, but this is a very uncertain number.

E. Abundances

Abundance determinations from the lines in the visible part of the spectrum are limited essentially to Ca and Na, although a few measurements of Ti^+, Fe^o, and K^o are known. The difficulty with Ca and Na is that we observe Ca^+ and Na^o, and it is likely that most of the Na is in the form of Na^+ and Ca in the form of Ca^{++}. If we know the ionization equilibrium and n_e, we can correct for this. The results obtained for the abundances after correction give essentially the same result for all ten "clouds" for which this procedure can be applied. The results are:

$$\frac{Na}{H} \simeq 0.7 \text{ to } 1 \times 10^{-6}$$

$$\frac{Ca}{H} \simeq 3 \text{ to } 5 \times 10^{-9}$$

This result may be compared with stellar atmosphere abundance determinations where $Na/N \simeq Ca/H \simeq 10^{-6}$. Thus, the sodium abundance agrees well with the stellar atmosphere determination, while the calcium abundance is a factor of 200-500 lower.

Recently, thanks to the ultraviolet observations (De Boer et al., 1972; Morton et al., 1973) results on abundances of other elements are also now known, although they are new and preliminary. The following results are probably correct to within a factor of three:

	Interstellar medium	Stellar atmosphere
Mg/H	3×10^{-6}	4×10^{-5}
Fe/H	2×10^{-6}	4×10^{-5}
C/H	3×10^{-4}	6×10^{-4}
N/H	3×10^{-4}	10^{-4}
O/H	3×10^{-4}	6×10^{-4}
Si/H	6×10^{-6}	4×10^{-5}

Thus, it appears that Mg, Si, and Fe are all a factor of ten to twenty less abundant in the interstellar medium than in stellar atmospheres, but not so underabundant as Ca. Carbon and oxygen may be slightly underabundant but the difference is probably not significant. Nitrogen appears to be overabundant in the interstellar medium, certainly with respect to C and O.

Why the abundances differ between the interstellar medium and the stellar atmosphere is not clear. Spitzer and Routly (1952) suggested that some of the gas condensed on dust grains, and this suggestion has been much repeated since. However, it can only really be justified if by a theory which will predict why certain elements only are depleted, and in the observed ratios and why these observed ratios are rather constant from place to place in the interstellar medium. Until then we must leave this as one of the many intriguing, but unsolved problems concerning the interstellar medium.

REFERENCES

Bély, O., van Regemorter, H., Ann.Rev.Astron.Astrophys. 8, 329, 1972.

Bergeron, J., Souffrin, S., Astron.Astrophys. 11, 40,1971.

Boksenberg, A., Kirkham, B., Towlson, W.A., Venis, T.E., Bates, B., Courts, G.R., Carson, P.P.D., Nature Phys. Sci. 240, 127, 1972.

Bridle, A.H., Nature 139, 246, 1969.

Brocklehurst, M., Mon.Not.Roy.Astron.Soc. 148, 417, 1970.

Burgess, A., Mon.Not.Roy.Astron.Soc. 118, 477, 1958.

Clark, B.G., Astrophys.J. 142, 1398, 1965.

Dalgarno, A., McCray, R.A., Ann.Rev.Astron.Astrophys. 10, 375, 1972.

Davies, R.G., Matthews, H.E., Pedlar, A., Nature Phys.Sci. 238, 101, 1972.

De Boer, K.S., Hoekstra, R., Van der Hucht, K.A., Kamperman, T.M., Lamers, H.J., Pottasch, S.R., Astron. Astrophys. 21, 447, 1972.

Dulk, G.A., Slee , O.B., Austr.J.Phys. 25, 429, 1972.

Dunham,Jr. T., Proc.Am.Phil.Soc. 81, 277, 1939.

Ellis, G.R.A., Hamilton, P.A., Astrophys.J. 143, 227, 1966.

Field, G.B., Interstellar Matter in Galaxies, Benjamin, New York, p. 183, 1962.

Field, G.B., Goldsmith, D.W., Habing, H.J., Astrophys.J. Lett. 155, 49, 1969.

Field, G.B., Steigman, G., Astrophys.J. 166, 59, 1971.

Gordon, M.A., Astrophys.J. 174, 361, 1972.

Goss, W.M., Astron.Astrophys. 18, 484, 1972.

Gottesmann, S.T., Gordon, M.A., Astrophys.J.Lett. 163, L93, 1970.

Habing, H.J., Bull.Astron.Inst.Netherl. 19, 421, 1968.

Hartmann, J., Astrophys.J. 19, 268, 1907.

Hayakawa, S., Nishimura, S., Takayanagi, K., Publ.Astron. Soc. Japan 13, 184, 1960.

Hjellming, R.M., Gordon, C.P., Gordon, K.J., Astron. Astrophys. 2, 202, 1969.

Hobbs, L.M.. Astrophys.J. 157, 135, 1969.

Hughes, M.P., Thompson, A.R., Colvin, R.A., Astrophys.J. Suppl. 23, 323, 1971.

Jackson, P.D., Kerr, F.J., Astrophys.J. 168, 29, 1971.

Johnson, H.M., Astrophys.J. 174, 591, 1972.

Kepner, M., Astron.Astrophys. 5, 444, 1970.

Marschall, L.A., Hobbs, L.M., Astrophys.J. 173, 43, 1972.

Monnet, G., Astron.Astrophys. 12, 379, 1971.

Morton, D.C., Drake, J.F., Jenkins, E.B., Rogerson, J.B., Spitzer, L., York, D.G., Astrophys.J., 1973.

Pikelner, S.B., Sov.Astron.A.J. 11, 737, 1967.

Plaskett, J.S., Pearce, J.A., Publ.Donn.Astrophys.Obs. Victoria 5, 167, 1933.

Pottasch, S.R., Astron.Astrophys. 17, 128, 1972a.

Pottasch, S.R., Astron.Astrophys. 20, 245, 1972b.

Radhakrishnan, V., Murray, J.D., Lockhart, P., Whittle, R.P.J., Astrophys.J. Suppl. 203, 15, 1972.

Reynolds, R.J., Roesler, F.L., Scherb, F., Astrophys.J. (submitted for publication), 1973.

Routly, P.M., Spitzer,Jr. L., Astrophys.J. 115, 227, 1952.

Sancisi, R., Klomp, M., Astron.Astrophys. 18, 329, 1972.

Seaton, M.J., Mon.Not.Roy.Astron.Soc. 120, 326, 1960.

Spitzer,Jr. L., Diffuse Matter in Space, Interscience Publishers, New York, 1968.

Strömgren, B., Astrophys.J. 108, 242, 1948.

White, R.E., Astrophys.J. (submitted for publication), 1973.

Witt, A.N., Johnson, M.W., Astrophys.J. (submitted for publication), 1973.

MOLECULES IN DENSE INTERSTELLAR CLOUDS

B.J. Robinson

C.S.I.R.O. Radiophysics Laboratory
Sydney, Australia

1. INTRODUCTION

Observations of complex interstellar molecules have excited much interest because:

(1) They reveal dense, cool clouds which are likely sites for star formation.

(2) The clouds are composed mainly of molecules and dust. They cannot be observed in the 21 cm line because all the hydrogen is molecular. They cannot be observed in the UV because the extinction is much too great. There is no radio continuum emission. However, IR emission from the dust can be observed.

(3) There is as yet no satisfactory theory to explain how complex organic molecules can form under conditions remote from those of terrestrial chemistry-pressures less than 10^{-15} atmospheres at temperatures of $50^{o}K$ or less.

(4) Some of the molecules found are basic building blocks for biological molecules.

G.R. Carruthers has already dealt with the UV observations of molecules. The subject of OH and H_2O masers has been dealt with by H.J. Habing and will not be discussed here. Nor will I deal with circumstellar molecules.

2. WHAT MOLECULES HAVE BEEN FOUND?

UV and radio observations have found a total of 36 molecules containing H, C, O, N, S, Si and isotopes of H, C, O, N and S. The molecules detected in the radio spectrum are listed in Table I. Many polyatomic organic molecules are observed with up to seven atoms: HC_3N, HCOOH, CH_2NH, CH_3OH, CH_3CN, $HCONH_2$, CH_3CHO and CH_3C_2H. Some of the molecules are transient and rare under laboratory conditions (e.g. OH, CN, CH_2S, CH_2NH). Some lines have not been identified and may also be produced by transient molecules; e.g. the line at 90.7 GHz might be produced by HNC.

3. WHAT FREQUENCIES ARE OBSERVED?

Ten years ago radio astronomers knew of one spectral line, the 21 cm hyperfine line of atomic hydrogen. Table I lists 106 transitions from molecules, most of them discovered since 1969. They span a frequency range from 0.8 to 170 GHz, the whole observable microwave and millimeter-wave spectrum.

Diatomic and linear molecules. Most of the transitions are rotational transitions, end-over-end rotation of the molecules. For diatomic and linear molecules the transition frequencies depend on the moment of inertia I of the molecule, the energy states being given by

$$E = (\hbar^2/2I)\cdot J(J+1) = hB\cdot J(J+1) \qquad J = 0,1,2,... \tag{1}$$

where the rotational constant B is inversely proportional to I. $\Delta J=1$ transitions are allowed, so that the transition frequencies are

$$\nu = \frac{1}{h}(E_{J+1} - E_J) = 2B(J+1) \tag{2}$$

Values of B range from 57.5 GHz for the light molecule CO to 4.5 GHz for the long and heavy molecule H-C≡C-C≡N. HCN, CS and OCS have intermediate values of B.

Asymmetric-top molecules. In molecules such as $H_2C=O$, $H_2C=S$, $H_2C=NH$, NH_2CHO and CH_3CHO the moments of intertia about the principal axes of the molecule are unequal, and this produces "asymmetry doublets" on the rotational levels. If the two largest moments of inertia are I_B and I_C and we define rotational constants $B \propto 1/I_B$ and

(1)	(2)	(3)	(4)	(5)	(6)	(7)	(8)	(9)	(10)	(11)	(12)	(13)	(14)	(15)	(16)
No. of atoms	Molecule		Transition	Frequency	Em. or abs.	Column density	Sources where detected								
	Name	Formula		GHz		mols cm^{-2}	Sgr B2	Sgr A	Or/A	W51	DR21	IRC +10216	W3	NGC 2264	Other
2	Hydroxyl	^{16}OH	$^2\Pi_{3/2}$, J=3/2, F=1-2	1.612231	A&E		*	*		*			*		≈ 200
			" " F=1-1	1.665401	A&E		*	*	*	*	*		*	*	"
			" " F=2-2	1.667358	A&E	10^{12}-10^{16}	*	*	*	*	*		*		"
			" " F=2-1	1.720533	A&E		*	*		*			*		>100
			$^2\Pi_{3/2}$, J=5/2, F=2-2	6.030739	E								*		1
			" " F=3-3	6.035085	E		*				*		*		5
			$^2\Pi_{3/2}$, J=7/2, F=4-4	13.441371	E								*		
			$^2\Pi_{1/2}$, J=1/2, F=0-1	4.660242	E		*		*						
			" " F=1-1	4.750656	E		*								
			" " F=1-0	4.765562	E		*						*		2
		^{18}OH	$^2\Pi_{3/2}$, J=3/2, F=1-1	1.63753	A		*	*							
			" " F=2-2	1.63948	A		*	*							
	Carbon monoxide	$^{12}C^{16}O$	J=1-0	115.2712	E	10^{17}-10^{19}	*	*	*	*		*	*		>20
		$^{13}C^{16}O$	"	110.2014	E		*	*		*		*			5
		$^{12}C^{18}O$	"	109.7822	E		*			*					
		$^{12}C^{17}O$	"	112.3593	E		*								

Table I: Interstellar molecules observed at radio frequencies (up to March 1973).
(continued next page)

(1)	(2)	(3)	(4)	(5)	(6)	(7)	(8)	(9)	(10)	(11)	(12)	(13)	(14)	(15)	(16)
2	Cyanogen	CN	N=1-0	113.491	E	10^{15}	*	*	*	*		*			
	Carbon mono-	$^{12}C^{32}S$	J=3-2	146.96916	E	10^{14}			*	*	*	*			
	sulphide	"	J=2-1	97.98101	E		*							*	≈3
		"	J=1-0	48.99100	E		*								≈5
		$^{12}C^{34}S$	J=2-1	96.41295	E		*								
		"	J=1-0	48.20695	E		*								
		$^{13}C^{32}S$	J=2-1	92.494	E		*								
		"	J=1-0	46.24747	E		*								
	Silicon	SiO	J=2-1	86.847	E	10^{13}	*		*						
	monoxide		J=3-2	130.2684	E	10^{13}	*								
3	Water	H_2O	$6_{16}-5_{23}$	22.23508	E	Maser	*		*	*	*		*		>25
	Hydrogen	$H^{12}C^{14}N$	J=1-0	88.63185	E	10^{15}	*	*	*	*	*	*	*	*	3
	cyanide	$H^{13}C^{14}N$	J=1-0	86.34005	E			*	*			*			
		$H^{13}C^{14}N$	J=2-1	172.6777	E				*						
		$H^{12}C^{15}N$	J=2-1	172.1081	E				*		*	*			
	Deuterium	$D^{12}C^{14}N$	J=1-0	72.4147	E	$4x10^{12}$			*						
	cyanide	$D^{12}C^{14}N$	J=2-1	144.8280	E				*						
	Carbonyl	OCS	J=9-8	109.4628	E	10^{14}	*								
	sulphide		J=6-5	72.9768	E		*								

Table I (continued)

(1)	(2)	(3)	(4)	(5)	(6)	(7)	(8)	(9)	(10)	(11)	(12)	(13)	(14)	(15)	(16)
3	Hydrogen sulphide	H_2S	$1_{10}-1_{01}$	168.76276	E	10^{14}	*		*	*	*		*	*	2
4	Ammonia	NH_3	J,K=1,1-1,1	23.69448	E	10^{16}	*	*	*	*	*		*		6
			2,2-2,2	23.72271	E		*	*		*					2
			3,3-3,3	23.87011	E		*	*		*					2
			4,4-4,4	24.12939	E		*								
			6,6-6,6	25.05604	E		*								
			2,1-2,1	23.09879	E		*								
			3,2-3.2	22.83417	E		*								
	Formaldehyde	$H_2{}^{12}C^{16}O$	$1_{10}-1_{11}$	4.829660	A	$10^{12}-10^{16}$	*	*	*	*	*		*		≈100
		(ortho)	$2_{11}-2_{12}$	14.48865	A		*	*		*					3
			$3_{12}-3_{13}$	28.97485	A		*								
			$5_{14}-5_{15}$	72.40909	E				*						
			$2_{12}-1_{11}$	140.83953	E			*	*	*			*		
			$2_{11}-1_{10}$	150.49836	E				*						
		(para-)	$2_{02}-1_{01}$	145.60297	E				*						
		(")	$1_{01}-0_{00}$	72.83797	E		*		*						
		$H_2{}^{13}C^{16}O$	$1_{10}-1_{11}$	4.593089	A		*	*		*					
		$H_2{}^{12}C^{18}O$	$1_{10}-1_{11}$	4.388797	A		*								
	Thio-formaldehyde	H_2CS	$1_{10}-1_{11}$	1.04648	A					*					
			$2_{11}-2_{12}$	3.13938	A	10^{16}	*	?							

Table I (continued)

(1)	(2)	(3)	(4)	(5)	(6)	(7)	(8)	(9)	(10)	(11)	(12)	(13)	(14)	(15)	(16)
4	Isocyanic acid	HNCO	$1_{01}-0_{00}$	21.9817	E	10^{14}	*								
			$4_{04}-3_{03}$	87.92545	E		*								
			$4_{13}-3_{12}$	88.23905	E		*								
			$5_{05}-4_{04}$	109.90585	E		*								
5	Cyanoacetylene	HC_3N	J=1-0, F=1-1	9.097036	E		*								
			F=2-1	9.098332	E	10^{16}	*								
			F=0-1	9.100279	E		*								
			J=2-1	18.196275	E		*								
			J=8-7	72.8	E		*								
			J=9-8	81.3	E		*								
			J=10-9	91.0	E		*								
			J=11-10	100.1	E		*	*	*	*					
	Formic acid	HCOOH	$1_{10}-1_{11}$	1.638805	E	10^{13}?	*								
	Methanimine	CH_2NH	$1_{10}-1_{11}$, F=0-I	5.28900			*								
			F=2-2	5.28982	E	10^{14}	*								
			F=2-1	5.29065			*								
			F=1-2	5.29085			*								
			F=1-1	5.29170			*								

Table I (continued)

(1)	(2)	(3)	(4)	(5)	(6)	(7)	(8)	(9)	(10)	(11)	(12)	(13)	(14)	(15)	(16)
6	Methancl	CH_3OH	1_1-1_1 (A)	0.834267	E		*	*							
			3_1-3_1 (A)	5.00532	E		*								
			4_2-4_1 (E_1)	24.93347	E	10^{16}			*						
			5_2-5_1 (E_1)	24.95908	E				*						
			6_2-6_1 (E_1)	25.01814	E				*						
			7_2-7_1 (E_1)	25.12488	E				*						
			8_2-8_1 (E_1)	25.29441	E				*						
			1_0-0_0 (A)	48.37260	E	10^{16}	*								
			1_0-0_0 (E)	48.37709	E		*								
			5_1-4_0 (E_2)	84.52121	E	10^{17}	*								
	Methylcyanide	CH_3CN	6_5-5_5	110.3307	E		*								
			6_4-5_4	110.3497	E		*								
			6_3-5_3	110.3645	E		*	?							
			6_1-5_1	110.3814	E		*	*							
			6_0-5_0	110.3835	E	10^{14}	*	*							

Table I (continued)

(1)	(2)	(3)	(4)		(5)	(6)	(7)	(o)	(9)	(10)	(11)	(12)	(13)	(14)	(15)	(16)
6	Formamide	$HCONH_2$	$1_{10}-1_{11}$,	F=1-1	1.538135	E		*	*							
				F=1-2	1.538693	E		*	*							
				F=2-1	1.539295	E		*								
				F=1-0	1.539570	E		*								
				F=2-2	1.539851	E	10^{15}	*	*							
				F=0-1	1.541018	E		*	*							
			$2_{11}-2_{12}$,	F=2-2	4.61714	E		*								
				F=3-3	4.61900	E		*								
				F=1-1	4.62001	E		*								
7	Acetaldehyde	CH_3CHO	$1_{10}-1_{11}$		1.065075	E	10^{14}	*	*							
			$2_{11}-2_{12}$		3.195167	E	10^{14}	*								
	Methylacetylene	CH_3C_2H	5_0-4_0		85.45729	E	?	*								
			5_3-4_3		85.44261	E		*								
?	U85.4	?	?		85.435	E		*								
?	U89.2	?	?		89.1890	E			*	*	*			*	*	I
?	U90.7	?	?		90.6639	E				*					*	
?	U144.9	?	?		144.8579											
?	U169.3	?	J=2-1?		169.3361	E				*						

Table I (continued)

$C \propto 1/I_C$, the doublet splitting is given in Table II.

Rotational Level	K-Doublet Transition Frequency	Typical Transitions observed (GHz)		
		H_2CO	NH_2CHO	CH_3CHO
J = 0	0	-	-	-
J = 1	B-C	4.830	1.539	1.065
J = 2	3(B-C)	14.489	4.619	3.195

Table II: Asymmetry doublet frequencies.

The asymmetry-doublet transitions thus span the decimetre- and centimetre-wave spectrum.

Other types of transition. Pure rotational and asymmetry-doublet transitions account for most of the lines in Table I. Other types of transitions observed include Λ-doubling for OH, the inversion spectrum of NH_3, and hindered internal rotation in CH_3OH. For an explanation of these consult Gordy and Cook (1970). For molecules containing nitrogen its nuclear spin produces a characteristic multiplet structure.

4. CONFIDENCE LEVEL OF THE IDENTIFICATIONS

(a) For most of the molecules in Table I several transitions are observed and there can be no doubt about the identification. Figure 1 shows 7 of the transitions observed in H_2CO.

(b) When a molecule contains a nitrogen atom the multiplet structure gives firm confirmation of the identification. Figure 2 shows the multiplet on which the identification of CH_2NH is based.

(c) When only one line can be observed for a molecule the detection of isotopic substitutions can be used to confirm the identification. For CO only the $J=1 \rightarrow 0$ transition has been seen, the $J=2 \rightarrow 1$ and higher transitions

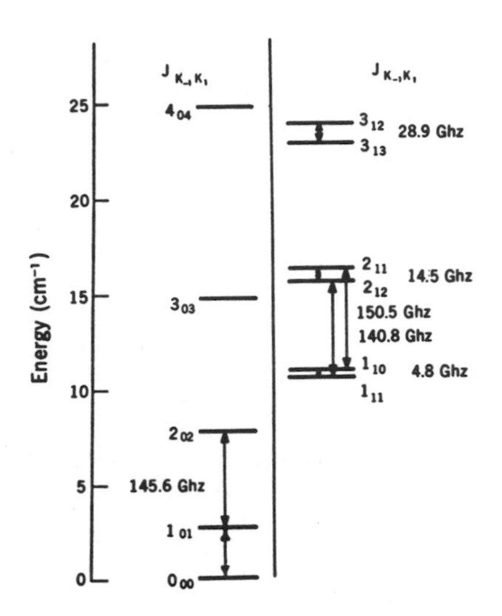

Fig. 1: Energy level diagram for the lower rotational levels of H_2CO. Levels are designated with the notation $J_{K_{-1}K_1}$ where J is the total angular momentum. K_{-1} and K_1 are angular momenta about the symmetry axes for the corresponding levels of a prolate and oblate symmetric top, respectively. The right-hand series of levels are those of orthoformaldehyde, and the left-hand series those of paraformaldehyde. Arrows show the transitions that have so far been observed from interstellar clouds. (From Rank et al., 1971)

lying above 200 GHz (see equation 2). However, the following isotopic substitutions of CO have been found:

$$^{12}C^{16}O \qquad \nu = 2B = 115.2712 \text{ GHz}$$

$$^{13}C^{16}O \qquad \nu = 2B = 110.2014 \text{ GHz}$$

$$^{12}C^{17}O \qquad \nu = 2B = 112.3593 \text{ GHz}$$

$$^{12}C^{18}O \qquad \nu = 2B = 109.7822 \text{ GHz}$$

(d) The identifications of H_2O, H_2S and HCOOH depend on the detection of a single line. However, the close agreement with the measured laboratory frequency to better than 1 part in 10^5 makes misidentification unlikely. No

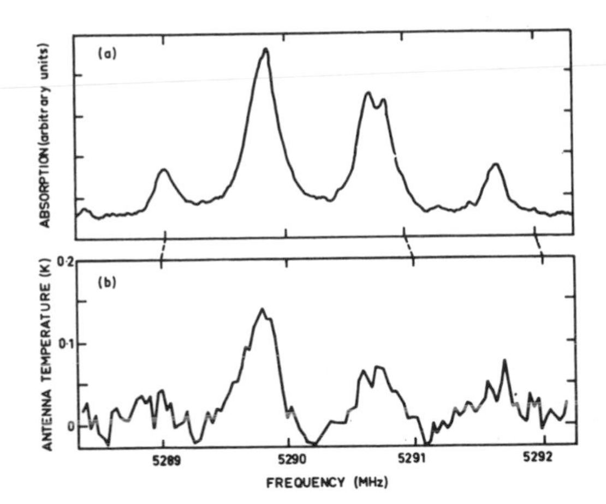

Fig. 2: (a) Laboratory spectrum of methanimine (CH_2NH) $1_{10} \leftarrow 1_{11}$ transition. (b) $1_{10} \rightarrow 1_{11}$ emission of methanimine observed in Sgr B2. The scale of rest frequencies is based on a radial velocity of 63 km sec^{-1}. (From Godfrey et al., 1973)

microwave spectrum of CN has been measured, and the assignment of the 113.491 GHz line to CN cannot be regarded as certain.

(e) Blending with other lines makes certain molecular lines difficult to measure. The $1_{10} - 1_{11}$ line of formic acid at 1.63881 GHz is confused with the broad absorption of Sgr A and Sgr B2 by lines of ^{18}OH at 1.63753 and 1.63948 GHz; confirmation of the detection of formic acid thus requires observation of other transitions, but attempts to find them have not been successful. Other molecular lines are confused with hydrogen recombination lines: the $1_{10} - 1_{11}$ line of H_2CS at 1.04648 GHz with H184α , the $2_{11} - 2_{12}$ line of $CHONH_2$ at 4.61900 GHz with H112α , the $3_1 - 3_1$ line of CH_3OH at 5.00532 GHz with H137β , the $1_{10} - 1_{11}$ line of HDCO at 5.346141 GHz with H134β ...

5. WHERE ARE THE MOLECULES FOUND?

(a) Widespread molecules: A number of molecules are found widely distributed in the galaxy, such as CO, OH, CS, HCN, H_2O, H_2S, NH_3 and H_2CO. These molecules are mainly associated with HII regions, and sometimes with

IR objects. CO, OH and H_2CO are also found in dark clouds.

(b) Simple molecules: An examination of Table I shows that diatomic and triatomic molecules are found in great abundance in a few objects, mainly HII regions. The number of molecules with up to four atoms in these objects are:

Sgr B2:	19 species	Sgr A :	9 species
Ori A :	16 "	IRC+10216 :	8 "
W 51 :	14 "	DR 21 :	7 "

(c) Complex molecules: Sgr B2 is the prime source of polyatomic organic molecules with up to seven atoms, such as HC_3N, HCOOH, CH_2NH, CH_3OH, CH_3CN, $HCONH_2$, CH_3CHO and CH_3C_2H. Some of these molecules are also found in Sgr A, while HC_3N and CH_3OH have been reported in Ori A. As we shall see later, not enough is known about the formation of complex molecules to enable us to understand why Sgr B2 is such a unique source.

6. SIZES OF MOLECULAR CLOUDS

The molecular clouds in Ori A and Sgr B2 have been mapped and found to be quite large. For Ori A Figure 3 shows the distribution of the 140 GHz emission from the $2_{12} - 1_{11}$ transition of H_2CO, as measured with the Kitt Peak 11 metre dish (beamwidth 55 arc sec). The size of the cloud is about 2 arc min, which corresponds to 1 parsec at a distance of 1.7 kpc. For Sgr B2 Figure 4 shows the distribution of the 9 GHz emission from the J=1-0 transition of HC_3N, as measured with the Parkes 64 metre dish (beamwidth 2.4 arc min). The molecular cloud has a size of about 5 arc min, which corresponds to 15 parsec at a distance of 10 kpc.

The available data on sizes of molecular clouds is collected in Table III. High resolution observations of OH and H_2CO absorption by lunar occultations have shown no evidence for structure less than a few minutes of arc. Similarly, aperture synthesis measurements on H_2CO absorption have not revealed any fine structure. There is a great need for high resolution measurements on other molecules, and we can expect some results to come from the prototype mm-wave interferometer at Bordeaux.

Within these molecular clouds are found the OH and H_2O masers, which are ultra-dense condensations with typical sizes of only a few A.U.

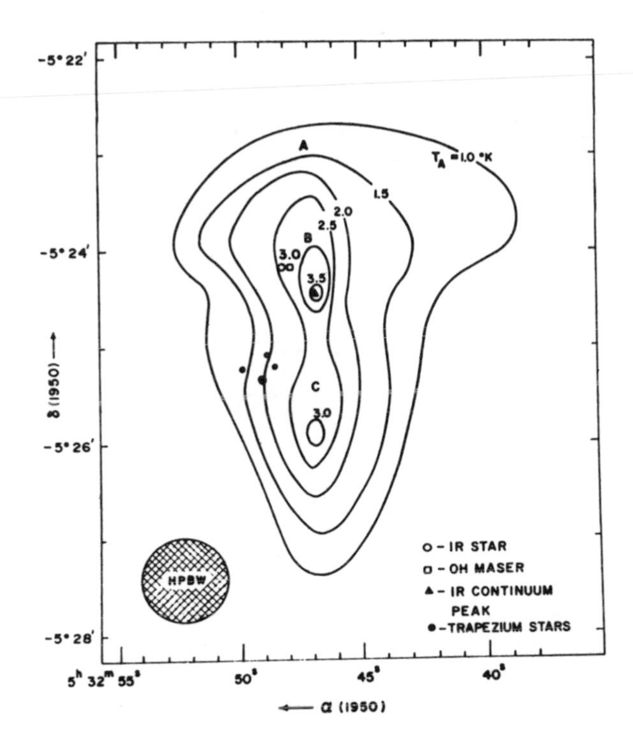

Fig. 3: Map of 140-GHz H_2CO emission in Ori A, shown in relation to the infrared star of Becklin and Neugebauer, the point source of OH emission and the continuum peak of the infrared nebula. (From Thaddeus et al., 1971)

7. TEMPERATURES OF MOLECULAR CLOUDS

(a) Rotational excitation temperatures. The brightness temperature T_b of a molecular cloud is given by

$$T_b = (T_{ex} - T_{BG})(1 - e^{-\tau}) \quad (3)$$

where T_{ex} is the rotational excitation temperature, T_{BG} is the continuum background temperature and τ is the optical depth of the line. All mm-lines are seen in emission, and so $T_{ex} > T_{BG}$.

For the $^{12}C^{16}O$ line at 115 GHz we know from observations of isotopic species that $\tau \gg 1$ in several sources

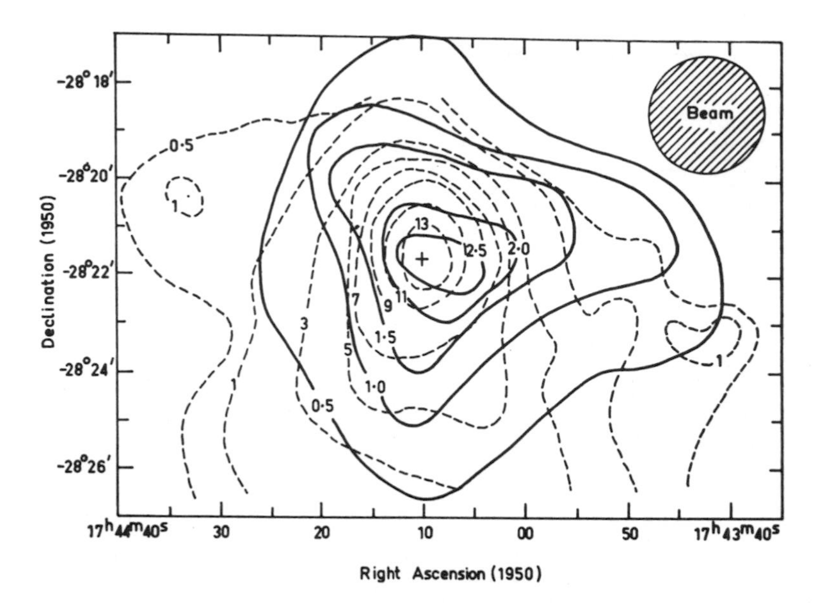

Fig. 4: The continuum radiation at wavelength 3.4 cm in the direction of Sgr B2 is shown by the broken contours. The distribution of HCCCN line emission is shown by the full contours, which give the integral under the profile in units of K MHz (1 K MHz = 2.54×10^{-26} W m^{-2} $sterad^{-1}$). Estimates of molecular column density $\int N dl$ (molecules cm^{-2}) may be obtained by multiplying by 5.3×10^{16}. (From McGee et al., 1973)

and so $T_b \doteqdot (T_{ex} - T_{BG})$, with $T_{BG} \doteqdot 2.7$ K. In Orion A the observed T_b is 45 K, while in Sgr B2 T_b is 15 K (with $\tau \simeq 90$). In dark clouds $\tau \geqslant 20$ and values of T_{ex} between 6 K and 18 K are found.

For the ammonia transitions at 25 GHz $\tau < 1$, but the relative intensities of the 1 - 1 and 2 - 2 lines can be used to determine an excitation temperature. In Sgr B2 the observed intensity ratios correspond to 20 K $< T_{ex}$ $<$ 80 K.

Values of $T_{ex} > 30$ K are also required to produce significant populations in the high J states observed in some molecules (e.g. J=6 in CH_3CN and J=9 in OCS).

(b) Kinetic temperatures. The rotational excitation is always a competition of interaction with the microwave radiation field (at rate $1/t_r$) and collisions with neutral molecules (mainly H_2) and with ions (at a rate $1/t_c$). This competition yields

SOURCE	MOLECULE	TECHNIQUE	RESOLUTION	STRUCTURE FOUND	REFERENCE
Sgr B2	OH	Occultation	3"	≈ 5'	Robinson et al.,1970
	H_2CO	Synthesis	40"	> 2·2'	Fomalont and Weliachew, 1973
	CO	11 m Dish	70"	≫ 5'	Solomon et al., 1972
	HC_3N	64 m Dish	2·5'	≈ 5'	McGee et al., 1973
	NH_3	25 m Dish	2·1'	> 8'	Cheung et al., 1969
Sgr A	OH	Occultation	6"	4'	Sandqvist, 1972
	H_2CO	Occultation	17"	4'	Sandqvist, 1972
	H_2CO	Synthesis	40"	5'	Fomalont and Weliachew, 1973
	CO	11 m Dish	70"	> 6'	Solomon et al., 1972
	NH_3	25 m Dish	1·9'	3'	Knowles and Cheung,1971
Ori A	CO	11 m Dish	70"	≫ 5'	Wilson et al., 1970
	CS	11 m Dish	55"	> 3'	Penzias et al., 1971
	H_2CO	11 m Dish	55"	2'	Thaddeus et al., 1971
W 49	H_2CO	Synthesis	40"	> 4'	Fomalont and Weliachew, 1973
	CO	11 m Dish	70"	2·5'	Scoville and Solomon, 1973
W 51	H_2CO	Synthesis	40"	> 2'	Fomalont and Weliachew, 1973
	CO	11 m Dish	70"	> 4'	Scoville and Solomon, 1973
	CS	11 m Dish	55"	3'	Penzias et al., 1971

Table III: Sizes of molecular clouds.

$$T_{ex} = \frac{t_r \cdot T_K + t_c \cdot T_{BG}}{t_r + t_c} \tag{4}$$

so that T_{ex} tends to T_K when $t_r \gg t_c$.

CO has a small dipole moment and so interacts weakly with the radiation field. We thus infer that the observed values of T_{ex} are close to the kinetic temperature of the gas. Thus, for the Orion A cloud $T_K \approx 40$ K and for Sgr B2 $T_K \approx 20$ K.

8. DENSITIES AND MASSES OF MOLECULAR CLOUDS

(a) Total densities in the range 10^4 to 10^6 molecules per cm^3 can be inferred from a number of arguments. The most direct argument is based on the column density $\int n\, dl$ which is proportional to the area under the line profile, $\int T_b(\nu)\, d\nu$, when $\tau \ll 1$. For the J=1→0 transition of a diatomic or linear molecule

$$\int_0^L n\, d\ell \doteqdot \left[\frac{3ck}{8\pi^3} \cdot \frac{g_0}{g_1} \cdot \frac{1}{|\mu|^2} \cdot \frac{1}{\nu^2} \cdot \int_0^\infty T_b(\nu) d\nu \right] \times \frac{k\, T_{ex}}{h\, B} \qquad (5)$$

where the quantity in square brackets is the column density of the J=0 level and the factor $k\, T_{ex}/hB$ (the rotational partition function) results from summing over all values of J. In equation (5) n is the number of molecules per cm^3, g_i is the statistical weight of level i, $|\mu|^2$ is the square of the dipole moment matrix element for the transition.

For CO in Sgr B2 we believe that $\tau \gg 1$ for $^{12}C^{16}O$ (which enables us to measure T_{ex}) but that $\tau < 1$ for $^{13}C^{16}O$ or $^{12}C^{18}O$. From the observed $\int T_b(\nu)\, d\nu$ for $^{13}C^{16}O$ in Sgr B2 we find $\int_0^L n\, dl \approx 2 \times 10^{17}$ molecules per cm^2. If $[H/^{13}C]$ had the terrestrial abundance ratio of 90 x 3000, and all the carbon atoms were tied up in CO molecules, the column density of H_2 molecules would be 3×10^{22} molecules/cm^2. This must be a lower limit to the column density of H_2 molecules. For a cloud thickness of, say, 10 parsec we thus infer a density $n_{H2} > 10^3\ cm^{-3}$.

(b) Rotational levels are excited by collisions with hydrogen molecules at a rate

$$1/t_c \approx n_{H_2}\, \sigma\, v \qquad (6a)$$

where σ is the collision cross section ($\approx 10^{-15} cm^{-2}$), v is the thermal velocity ($\approx 5 \times 10^4$ cm sec^{-1}). Equilibration with the background radiation T_{BG} takes place at a rate

$$\frac{1}{t_r} = \frac{64\, \pi^4}{3hc^3} \cdot \frac{\nu^3\, |\mu|^2}{[1 - \exp(-h\nu/kT_{BG})]} \qquad (6b)$$

Thus,

$$\frac{t_r}{t_c} \propto \frac{n_{H_2}}{\nu^3 \ |\mu|^2} \cdot \left[1 - \exp(-h\nu/kT_{BG})\right] \qquad (6c)$$

When $t_r > t_c$ equation (4) shows that T_{ex} will approach T_K. For $^{12}C^{16}O$ T_{ex} is observed to be in the range 20 K to 40 K, very much greater than T_{BG} = 2.7 K, and thus $t_r > t_c$. If we substitute μ_{CO} = 0.1 debye in equation (6c) we find that $n_{H_2} > 10^4$ cm^{-3}.

For mm-wave transitions in molecules such as CS, HCN, HNCO, H_2CO or CH_3CN with large dipole moments the density of H_2 molecules required to produce collisional excitation is much larger, of the order of 10^6 molecules per cm^3.

(c) Lower densities would be required if the molecules were excited by collisions with electrons. There are two reasons for believing that this is unlikely.

Firstly, it is believed that the dense molecular clouds contain sufficient dust to screen out UV photons which would dissociate the molecules. There would then not be sufficient UV photons to ionize the gas. Also, most molecules will dissociate at an energy well below their ionization potential. Low energy cosmic rays would produce ionization, but at too low a level to be important for rotational excitation.

Secondly, for electron collisions $t_c \propto 1/|\mu|^2$, and so t_r/t_c is independent of $|\mu|^2$. We would thus expect molecules such as CO (μ = 0.1 debye) and CS (μ = 2.0 debye) to have the same values of T_{ex}. But in Orion A the observed values of T_{ex} are 45 K for CO and 9 K for CS. This strongly suggests that the collisional excitation must be by neutral molecules.

(d) The molecular clouds observed at mm-wavelengths are much denser than those observed at UV wavelengths (see G.R. Carruthers' lecture). For $^{12}C^{16}O$ the radio observations give column densities of 10^{19} cm^{-2} or higher, while the UV observations give column densities of 10^{15} cm^{-2}. For CN, radio observations give 10^{15} cm^{-2}, while UV observations give 10^{12} cm^{-2}. The clouds observed in the UV have visual extinction A_v of the order of 1 magnitude. We infer that the denser clouds contain much more dust, and that A_v might be as large as 200 magnitudes.

(e) If $n_{H2} > 10^4$ cm^{-3} the dense molecular clouds must be very massive. For the Orion cloud (radius $\approx 1.5 \times 10^{18}$ cm) the mass will exceed 200 $M_\odot$. For the Sgr B2 cloud (radius $\approx 1.5 \times 10^{19}$ cm) the mass will exceed 2×10^5 $M_\odot$. As we saw in section 8 (b) the values of n_{H2} may be as high as 10^6 cm^{-3}, and the masses correspondingly larger.

Such clouds must be gravitationally unstable. The critical Jeans' Mass M_J is given by

$$\frac{M_J}{M_\odot} = 32\ T_K^{3/2}\ /\ n_{H_2}^{\frac{1}{2}}$$

For $T_K = 20$ K and $n_{H2} > 10^4$ cm^{-3}, $M_J < 30\ M_\odot$. The collapse time is inversely proportional to $(n_{H2})^{1/2}$; for $n_{H2} = 10^4$ cm^{-3} the collapse time is 3×10^5 years, while for $n_{H2} = 10^6$ cm^{-3} it reduces to 3×10^4 years.

9. ISOTOPIC SPECIES

Table I shows that many isotopic species can be detected. At radio wavelengths observations of ^{18}OH, $^{13}C^{16}O$, $^{12}C^{17}O$, $^{12}C^{18}O$, $^{13}C^{32}S$, $^{12}C^{34}S$, $H^{13}C^{14}N$, $H^{12}C^{15}N$, $D^{12}C^{14}N$, $H_2{}^{13}C^{16}O$ and $H_2{}^{12}C^{18}O$ have promised information on the abundances of ($^{12}C/^{13}C$), ($^{16}O/^{17}O$), ($^{16}O/^{18}O$), ($^{14}N/^{15}N$), ($^{32}S/^{34}S$) and (H/D). However, the optical depth of the common isotopic species has usually turned out to be high and of uncertain value, so that the ratios of the observed intensities give lower limits to the abundance ratios. Figure 5 shows that the intensitv ratio of $^{12}C^{16}O/^{13}C^{16}O$ for Sgr A varies markedly across the line profile: this is interpreted as the result of the varying optical depth of $^{12}C^{16}O$ across the profile.

For the optically thin lines of $^{13}C^{16}O$ and $^{12}C^{18}O$, and of $H_2{}^{13}C^{16}O$ and $H_2{}^{12}C^{18}O$, the ratio $(^{13}C/^{12}C)/(^{18}O/^{16}O)$ in Sgr B2 is 1.8 times the terrestrial ratio of 490/89=5.5. Nuclear burning in stars is unlikely to effect the ($^{18}O/^{16}O$) ratio, so that ($^{13}C/^{12}C$) may be enriched in the galactic centre by the C-N-O cycle in massive stars.

In Orion A the observations of $H^{12}C^{14}N$ and $H^{12}C^{15}N$ give $(^{14}N/^{15}N) \doteqdot 230$, in good agreement with the terrestrial abundance of 170 : 1. $H^{13}C^{14}N$ observations show

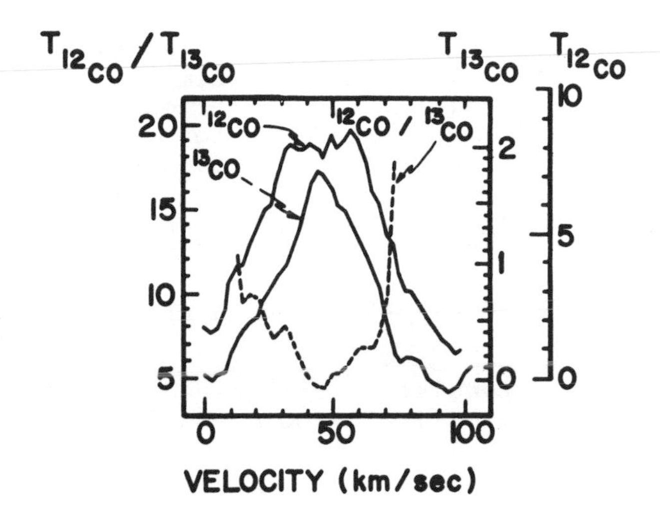

Fig. 5: Spectra of ^{12}CO and ^{13}CO near the galactic centre. The vertical scale is antenna temperature for the Kitt Peak 11 m dish. Also shown is $T_{^{12}CO}/T_{^{13}CO}$ for that portion of the spectra where $T_{^{13}CO}$ is higher than 3σ. (From Solomon et al., 1972)

that ($^{12}C/^{13}C$) in Orion cannot differ greatly from the terrestrial abundance.

$J = 1 \rightarrow 0$ and $2 \rightarrow 1$ transitions of DCN have been observed in Orion. Comparison of the intensity with that of other isotopes shows that for $J = 1 \rightarrow 0$ (D/H) = 0.7 ($^{13}C/^{12}C$ and for $J = 2 \rightarrow 1$ (D/H) = 1.6 ($^{15}N/^{14}N$). If the carbon and nitrogen abundances are close to their terrestrial values, (D/H) $\doteqdot 6 \times 10^{-3}$. This is 40 times greater than the value in terrestrial water and 80 times the limit obtained for atomic deuterium in the direction of Cas A. Nucleosynthesis in the Big Bang would produce (D/H) of only 10^{-5} (Wagoner, 1973), and any cycling through stars can only reduce (D/H). The surprisingly high density of DCN in Orion is attributed by Solomon and Woolf (1973) to enrichment of DCN by chemical fractionation. On the surface of low temperature dust grains the exchange reaction

$$HD + HCN \rightarrow H_2 + DCN$$

enriches the DCN abundance. For an atomic (D/H) abundance of about 10^{-5} a grain temperature of 80 K would produce (DCN/HCN) of 6×10^{-3}. Chemical fractionation will operate to enrich all molecular species containing deuterium relative to the corresponding species containing hydrogen.

10. FORMATION OF MOLECULES

The discovery of polyatomic molecules in interstellar clouds came as a great surprise to astronomers. It was known that diatomic molecules could be formed during atomic collisions, but this process is much too slow to build up larger molecules in successive collisions.

(a) Formation in stellar envelopes. One suggestion that has been proposed is that the molecules are formed in the atmospheres of cool stars and then expelled into the interstellar medium. This now seems unlikely for several reasons:

(i) The lifetime of molecules in unshielded regions of interstellar space is typically 10-100 years. In 100 years a molecule moving at 10 km sec^{-1} would travel only 10^{-3} parsec from the parent star.

(ii) The masses of molecular clouds are as high as 10^4 or 10^5 $M_\odot$, so very many stars would have to be expelling molecules.

(iii) The stars whose atmospheres are cool enough to form molecules have values of $(^{12}C/^{13}C)$ of less than 10, while in the molecular clouds the $(^{12}C/^{13}C)$ ratio lies in the range 50 to 100.

(iv) If molecules formed in stellar envelopes with $T \geqslant 1000$ K no enrichment of deuterated species would occur and we would expect (DCN/HCN) $\doteqdot 10^{-5}$.

(b) Formation inside molecular clouds. It is currently believed that the molecules are formed inside the molecular clouds. With typical temperatures of 50 K or less, only exothermic reactions need be considered. Possible gas-phase reactions are radiative association of ground-state atoms and chemical exchange reactions. Alternatively the molecules might be formed on the surfaces of dust grains, and then freed by evaporation or by decomposition of the grains.

(i) Gas phase reactions: Solomon and Klemperer (1972) have been able to account quantitatively for Herbig's observations of diatomic molecules in the direction of ζ Ophiuci. They assume a density of hydrogen atoms of 50 cm^{-3} and a temperature of 20 K. The principal reaction is the two-body radiative association of C^+ and H to form CH^+. CH can then be formed by recombination of CH^+ and an electron. Other molecules can then be formed by chemical exchange reactions, such as

$$CH^+ + O \rightarrow CO + H^+ \qquad CH + O \rightarrow CO + H$$

$$CH^+ + N \rightarrow CN + H^+ \qquad CH + N \rightarrow CN + H$$

$$CH^+ + C \rightarrow C_2^+ + H \qquad CH + C \rightarrow C_2 + H$$

$$CH + C^+ \rightarrow C_2^+ + H$$

This process will not be effective in the denser clouds observed at radio wavelengths because the shielding by dust will prevent the ionization of C, and the hydrogen will be mainly molecular. Thus, the basic reaction $C^+ + H \rightarrow CH^+ + h\nu$ will be rare.

(ii) Formation on grains: An atom colliding with a dust grain has a high probability of sticking, and then can spend a long time migrating on the surface before it evaporates off. During the migration it can combine with other atoms or radicals on the surface to form molecules. Formation on grains has been discussed by Watson and Salpeter (1972a, 1972b). The process is critically dependent on the temperature of the grains: if the temperature is too high the atoms or radicals will not stick for long enough to form molecules; if the temperature is too low the molecules will form but will stay trapped on the surface of the grain. For the formation of H_2 on grains their temperature must be above 13 K.

The molecules on the grain surface might be released if the grains are bombarded by energetic particles or photons, or if they are heated by shock waves.

(c) Destruction process. The main processes which remove molecules from the clouds are

(i) Photodissociation: The lifetime of an unshielded molecule in the interstellar UV field is typically less than 100 years. In the dense clouds it is believed that dust screens out the UV photons. What the photo-dissociation lifetimes are is uncertain since the extinction is not known. But Figure 6 shows that for a visual extinction A_v of only 4 magnitudes the lifetimes of many molecules are 10^6 to 10^7 years. CO has the highest dissociation energy of any molecule and has a very long lifetime, so its great abundance is not surprising.

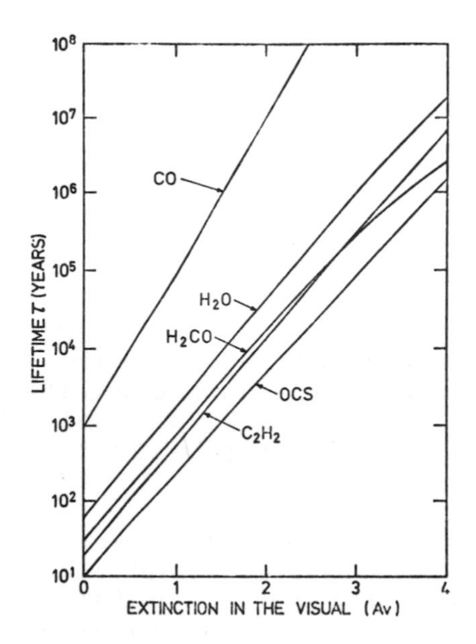

Fig. 6: Lifetime of interstellar molecules as a function of the visual extinction by dust, expressed in magnitudes. (From Stief, 1971)

(ii) Adsorption on dust grains: In the dense molecular clouds the molecules will stick to the grain surfaces within 100 years. How long they take to evaporate again depends critically on the grain temperature, as discussed above.

(iii) Exchange reactions: On grain surfaces at low temperatures exchange reactions will enrich some molecular species at the expense of others, as in the enrichment of DCN relative to HCN.

(d) Relative abundances of molecules. The column densities measured for different molecules are listed in Table I, and give some measure of their relative abundances. It is striking that polyatomic molecules such as HC_3N, CH_3OH and $HCONH_2$ are as abundant as the simple molecules CN or HCN. This could be a valuable clue to the processes which form the molecules.

In Table IV molecules have been grouped in simple families. Those that are underlined by a continuous line have been found in Sgr B2; those with a dashed underlining have been searched for but not found; the

$H-C\equiv X$	$H_2C=X$	H_3C-X	$H-C\equiv C-X$	$\mathrm{O{=}C(H)}-X$
$H-C\equiv N$	$H_2C=O$	H_3C-OH	$H-C\equiv C-OH$	$\mathrm{O{=}C(H)}-OH$
$H-C=O$	$H_2C=S$	H_3C-CN	$H-C\equiv C-CN$	$\mathrm{O{=}C(H)}-NH_2$
	$H_2C=NH$	$H_3C-C(=O)H$	$H-C\equiv C-NH_2$	$\mathrm{O{=}C(H)}-CH_3$
	$H_2C=CO$	$H_3C-C\equiv C-H$	$H-C\equiv C-CH_3$	$\mathrm{O{=}C(H)}-CN$
	$H_2C=NOH$	H_3C-NH_2	$H-C\equiv C-C(=O)H$	$\mathrm{O{=}C(H)}-C\equiv C-H$
		H_3C-NO		

Table IV: Organic molecules in Sgr B2.

remainder have not yet been searched for. We note that short-lived molecules such as $H_2C=S$ and $H_2C=NH$ have been found but the stable molecule $H_2C=CO$ has not. Clearly the dense clouds are far from chemical equilibrium. Another example of a transient molecule is HNC, which has been suggested as an identification for the line at 90.7 GHz. As noted earlier, the other unidentified lines may also be associated with transient species.

11. BIOLOGICAL IMPLICATIONS

Molecules such as HCN, H_2O, NH_3 and H_2CO, which abound in the interstellar gas, have been used by many workers as primordial planetary atmospheres in experiments that have synthesized organic compounds such as glycine, alanine, aspartic acid, glutanic acid, adenine, ribose, ... A search for the microwave spectra of such molecules in interstellar clouds is one of the most challenging problems of the next decade.

Organic evolution can clearly proceed much further than expected in the harsh, high vacuum conditions of interstellar space. Cyanoacetylene has been proposed many times as an intermediate in the formation of aspartic acid and cytosine. Formamide, formic acid and methanimine are other interstellar molecules that can be classed as prebiotic. Formic acid (HCOOH) may well react with methanimine (H_2CNH) to form glycine (NH_2CH_2COOH) - the simplest amino acid.

It is thus likely that organic evolution could be well advanced before gas clouds condense to form stars and planets. Although polyatomic molecules may not survive the heat of a young planet, the processes of molecular synthesis are likely to be repeated at a much faster rate in the denser, warmer primordial atmosphere of a planet. Also meteorites are certain to bring down some of the protoplanetary material from the parent gas cloud.

REFERENCES

A complete bibliography of papers on interstellar molecules from 1969-1972 is contained in the 1973 I.A.U. Reports on Astronomy.

Review papers:

Rank, D.M., Townes, C.H., and Welch, W.J., Science 174, 1083, 1971.

Solomon, P.M., Physics Today 26, 32, 1973.

Turner, B.E., Scientific American 228, 50, 1973.

References mentioned in text:

Cheung, A.C., Rank, D.M., Townes, C.H., Knowles, S.H., and Sullivan, W.T., Astrophys.J. 157, L13, 1969.

Fomalont, E.B., and Weliachew, L., Astrophys.J. 181, 781, 1973.

Godfrey, P.D., Brown, R.D., Robinson, B.J., and Sinclair, M.W., Astrophys.Lett. 13, 119, 1973.

Gordy, W., and Cook, R.L., Microwave Molecular Spectra, John Wiley, 1970.

Knowles, S.H., and Cheung, A.C., Astrophys.J. 164, L19, 1971.

McGee, R.X., Newton, L.M., Batchelor, R.A., and Kerr, A.R., Astrophys.Lett. 13, 25, 1973.

Penzias, A.A., Solomon, P.M., Wilson, R.W., and Jefferts, K.B., Astrophys.J. 168, L53, 1971.

Robinson, B.J., Goss, W.M., and Manchester, R.N., unpublished data, 1970.

Sandqvist, Aa., Astron.Astrophys.Suppl. 9, 391, 1972.

Scoville, N.Z., and Solomon, P.M., Astrophys.J. 180, 31, 1973.

Solomon, P.M., and Klemperer, W., Astrophys.J. 178, 389, 1972.

Solomon, P.M., Scoville, N.Z., Jefferts, K.B., Penzias, A.A., and Wilson, R.W., Astrophys.J. 178, 125, 1972.

Solomon, P.M., and Woolf, N.J., Astrophys.J. 180, L89, 1973.

Stief, L.J., Charlottesville Symposium on Interstellar Molecules, Ed. M.A. Gordon, John Wiley and Sons, in press, 1971.

Thaddeus, P., Wilson, R.W., Kutner, M., Penzias, A.A., and Jefferts, K.B., Astrophys. J. 168, L59, 1971.

Wagoner, R.V., Astrophys. J. 179, 343, 1973.

Watson, W.D., and Salpeter, E.E., Astrophys. J. 174, 321, 1972a.

Watson, W.D., and Salpeter, E.E., Astrophys.J. 175, 659, 1972b.

Wilson, R.W., Jefferts, K.B., and Penzias, A.A., Astrophys. J. 161, L43, 1970.

SOURCES FROM WHICH FIGURES HAVE BEEN TAKEN

Fig. 1 Figure 9 from Rank, Townes and Welch, Science, 174, 1083, 1971.

Fig. 2 Figure 1 from Godfrey, Brown, Robinson and Sinclair, Astrophys.Lett. 13, 119, 1973.

Fig. 3 Figure 2 from Thaddeus, Wilson, Kutner, Penzias, and Jefferts, Astrophys. J. 168, L59, 1971.

Fig. 4 Figure 3 from McGee, Newton, Batchelor and Kerr, Astrophys.Lett. 13, 25, 1973.

Fig. 5 Figure 3(b) from Solomon, Scoville, Jefferts, Penzias and Wilson, Astrophys. J. 178, 125,1972.

Fig. 6 From Stief, L.J., Charlottesville Symposium on Interstellar Molecules, Ed. M.A. Gordon, John Wiley and Sons, 1971.

INTERSTELLAR EXTINCTION IN THE ULTRAVIOLET FROM CELESCOPE OBSERVATIONS

K. Haramundanis

Smithsonian Astrophysical Observatory
Cambridge, Mass., U.S.A.

Celescope observations of 550 stars made in four broad ultraviolet passbands (1500 to 2600Å) were used to derive intrinsic ultraviolet colors for main sequence stars (Haramundanis and Payne-Gaposchkin, 1973). Color excesses, $E(Ui-V) = (Ui-V) - (Ui-V)_0$, are qualitatively in agreement with those obtained by the Wisconsin Experiment (Bless and Savage, 1972) for stars common to the two experiments. Excluding emission stars, binaries and supergiants whose ultraviolet fluxes cannot be compared with our intrinsic colors, we have verified that extinction in the ultraviolet is wavelength dependent. Our results further indicate that the slope of the reddening line differs with galactic longitude in the U2 (λ_{eff}=2308Å) and U3 (λ_{eff}=1621Å) passbands. The ratios E(U3-V)/E(B-V) for 185 stars selected with $E(B-V) \geq .14$ and averaged within regions of ten square degrees along the galactic equator are shown in the following fig. Stars range in spectral class from O4 to A0, with 50 % in the range from B0 to B3, and in visual magnitude from $5.40 \leq V \leq 10.12$.

The averaged ratios were tested for significant deviation from the mean, $\langle x \rangle = 4.33$, with the statistic

$$z = \frac{\bar{x}_i - \langle x \rangle}{\sigma / \sqrt{n}}$$

assuming a normal sampling distribution; they are significant at the 5 % level. The probability of obtaining the observed distribution by chance, deduced from a run test of these same means, is .005.

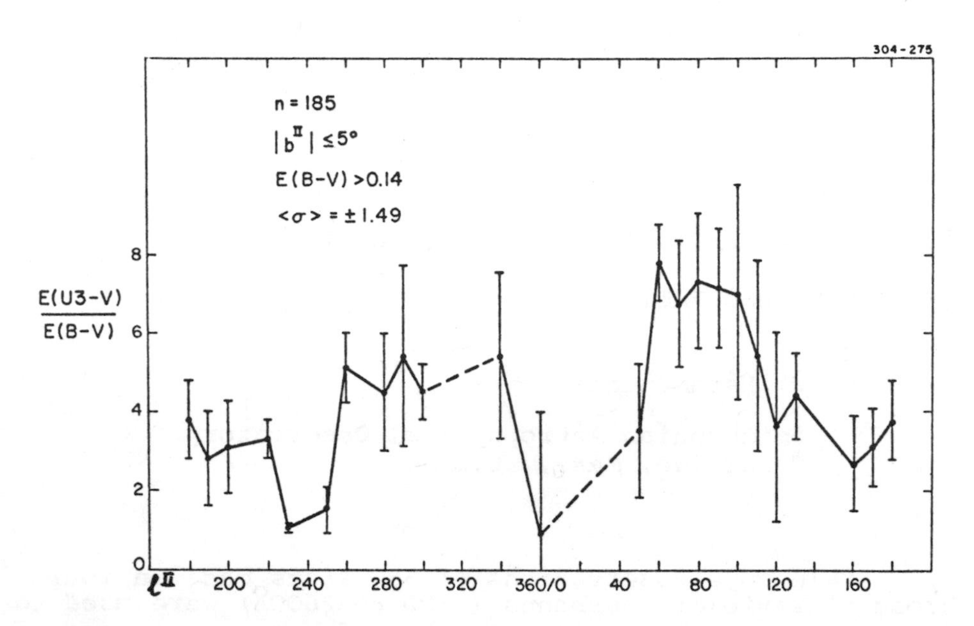

Error bars in the figure represent 1 rms deviation; the average deviation is ± 1.49. A dashed line indicates that the experiment made several exposures in the direction indicated but observed no objects. +)

REFERENCES

Bless, R.C., and Savage, B.D., The Scientific Results from the Orbiting Astronomical Observatory (OAO-2). NASA SP-310, p.175, 1972.

Haramundanis, K., and Payne-Gaposchkin, C., Astron. J., Vol. 78, June, 1973.

+) This research was supported by Grant NGR 09-015-200 from the National Aeronautics and Space Administration.

IDEAS ABOUT THE STUDY OF DUST

C.D. Andriesse

Kapteijn Astronomical Institute, Groningen,
The Netherlands

Compared to what is known about atoms and molecules, the actual knowledge on interstellar dust, as presented in the monographs by Greenberg (1968) and Wickramasinghe and Nandy (1972), is poor. In this course the points of view by van de Hulst, Woolf and Carruthers illustrate the lack of a well-established description, as is possible for e.g. the hydrogen atom and, to a lesser extent, the ammonia molecule. Agreement seems to exist about the general idea that a dust particle is a tiny piece of solid material, with some atomic lattice and bonding electrons. But questions like: how large are dust particles, what is their chemical composition, how do they scatter and absorb light, are answered differently when different spectral features are discussed. This incoherence in description is unsatisfactory, despite arguments that the nature of dust particles might depend on the type of stellar atmospheres where they possibly are formed.

Why do we know so much about atoms and molecules, so that they are useful probes to study conditions in interstellar space? Because their properties are studied in the laboratory with quantum theory as a guide. So why not start asking what can be said about tiny particles in general from experience in the laboratory? The first answer from physicists and chemists is, that such particles behave very different from macroscopic species, apparently because the numbers of atoms at the surface and in the bulk are comparable. The whole impressive theory for solid state lattices of infinite range and with long-range order (Kittel, 1963) does no longer apply.

Instead one has to treat a problem as badly defined as the surface is, possibly helped somewhat by the approximate description of systems with short-range order like glasses or liquids (Temperley et al., 1968). Furthermore one can try to understand at least qualitatively what is observed for such particles. Atoms at irregular surfaces quite often have unsaturated bonds, which are responsible for the observed strong chemisorption and light scattering. The latter property has already been pointed out to astronomers by Platt for the case of radical macromolecules (Platt, 1956). Quite often also characteristic vibrational modes of the bulk material are heavily damped and shifted to lower frequency, or disappear entirely when surface modes appear. An example of size influence on the magnetic field within a particle is the disappearance of the characteristic bulk Mössbauer spectrum of 6 lines in Fe_2O_3 when this is finely powdered. There are many other observations of this kind to warn for the use of bulk optical, vibrational and magnetic properties in the description of tiny particles. So the first answer you get is somewhat discouraging.

However, there is more to say. Combining concepts as quasi-electron bands and quasi-phonons, which are applicable in disordered systems (Bonch-Bruevich, 1968), with surface phenomena caused by radical electrons, it is possible to investigate some heuristic models for dust particles. Optical properties of such models should account for the conspicuous absorption band at 2200 Å (Bless and Savage, 1972), and some weaker bands in the visible (Herbig, 1966). Donn (1968) has drawn the attention to resonances for π electrons in benzene rings, which even might match all the observed energy levels (Johnson and Castro, 1971). Clearly this contains hints to the likely electron band structure. Furthermore the pure scattering for wavelengths below 1600 Å (Witt and Lillie, 1971) has to be explained, which seems entirely possible by radical electrons (Andriesse and de Vries). Vibrational properties of the lattice of such particle models should account for near-infrared resonances as described by Woolf. However, the lower the energy involved, the more uncertain becomes an identification, e.g. of the 9.7 μm band with the bulk resonance of a Si-O bond. In general one can say that the lowest possible optically active mode in small lattice of light atoms is somewhere around 30μm. Surface modes will critically depend on the shape of the surface and might well become irrelevant when a distribution over particle sizes and surface shapes is considered. The specific heat, which is linked mostly to the acoustical modes, is probably anomalously low, as in small particles these modes are most depressed. Debye temperatures might be in the range of those for "soft" liquids like Ar, so around 100 K. Using such ideas one easily can understand the sharp decrease in energy emission by dust particles in HII regions between 100 and 350 μm (Andriesse and Olthof, 1973). Magnetic properties should account for the observed polarization of starlight. What one needs is a directional preference for the electron spins involved, which evidently is the case for the benzene ring structure mentioned.

So it appears that observations give a number of keys for a reasonable particle model. When in addition to the model something can be derived for size distribution functions (given by the dynamical equilibrium between accretion and destruction processes), it seems entirely feasable to explain the observations with one coherent theory. Laboratory research is important to check the model(s) for those properties that cannot be directly observed. These include e.g. the inelasticity of 1-eV proton collisions, which might heat the dust particles in HII regions, (Spitzer, 1968) and electron fluorescence from electrons (Becker, 1969), which might explain the relatively high albedo in the visible.

REFERENCES

Andriesse, C.D., and Olthof H., submitted to Astron. Astrophys., 1973.

Andriesse, C.D., and de Vries, J., in preparation.

Becker, R.S., Theory and interpretation of fluorescence and phosphorescence, Wiley, 1969.

Bless, R.C. and Savage, B.D., Astrophys.J.171, 293,1972.

Bonch-Bruevich, V.L., Theory of condensed matter, p.989, IAEA Vienna, 1968.

Donn, B., Astrophys.J.152, L 129, 1968.

Greenberg, J.M., Stars and stellar systems VII, Ch.6, Chicago, 1968.

Herbig, G.H., Z. Astrophysik, 64, 512, 1966.

Johnson, F.M., and Castro, C.E., quoted by Heiles, C., Ann.Rev.Astron.Astrophys. p.294, 1971.

Kittel, C., Quantum theory of solids, Wiley, 1963.

Platt, J.R., Astrophys.J. 123, 486, 1956.

Spitzer, L., Diffuse matter in space, Ch. 4.5, Wiley, 1968.

Temperley, H.N.V., Rowlinson, J.S., and Ruhrsbrooke, G.S., Physics of simple liquids, North Holland, 1968.

Wickramashinghe, N.C., and Nandy, K., Rep. Progress Phys. 35, 157, 1972.

Witt, A.N., and Lillie, C.F., NASA SP-310, 199, 1971.

LARGE-SCALE DISTRIBUTION OF INTERSTELLAR MATTER IN THE GALAXY

J. Lequeux

Département de Radioastronomie
Observatoire de Paris, Meudon, France

INTRODUCTION

Studying the distribution of interstellar matter in our galaxy is a difficult task; the main difficulty arises from the absence of distance criteria for interstellar matter independent of kinematics, except when there is an obvious association of gas with stars for which a photometric distance can be obtained. Though the main motion in the galaxy is rotation, there are also random motions and systematic small- or large-scale non-circular motions: the differential rotation theory is thus only a first-order approximation, which can be used to obtain a gross overall picture of the spiral structure, but it gives relatively poor results for the solar vicinity, where fortunately optical observations are of great help. Attempts have been made for trying to detect non-circular motions to compare them with the predictions of the gravity-wave theory of spiral structure, and to have a better idea of the actual distribution of interstellar matter. This is certainly one of the most promising directions for the future, but much remains to be done in this field.

In these lectures, I have decided to study separately the distribution of interstellar matter in the vicinity of the sun, where optical observations are of major importance, and where observations at medium and high galactic latitudes provide information with little or no equivalent at large distances. Our only information on magnetic field and density of cosmic rays, and the bulk of our information on dust/gas ratio and chemical composition of the interstellar matter, comes from this region.

I. THE VICINITY OF THE SUN

The limits of this region are of course very arbitrary, and essentially observational. Available optical data generally do not extend beyond 4 kpc; those stars for which we possess detailed optical and UV observations of interstellar absorption lines are generally at less than 0.5 kpc. Most of the information we gather from pulsars or high-latitude objects refer to distances less than 1 kpc.

We first study the spiral structure close to the sun. Its complexity - and the fact that it is rather badly known, will serve as a warning when studying the gas density, the magnetic field, etc.

1. The spiral structure close to the sun

Spiral structure is defined by spiral tracers, which are mostly young stars or stellar clusters, HII regions and HI clouds and interstellar dust. A difficulty for studying spiral structure is the existence of a segregation between the various spiral tracers. In external galaxies, HII regions are the main arm tracers; dust lanes are generally in the inside of the arms (Lynds, 1970). Young stellar associations (type earlier than B3, age $\leqslant 10^7$ years) define arms close to the sun but older associations are scattered more uniformly (Becker and Fenkart, 1970). An interesting example of such a population segregation has been given in M33 by Courtès and Dubout-Crillon (1971) who have shown that most HII regions of the southern arm are concentrated along a well defined line, whilst blue stars, associations and red supergiants are distributed up to 800 pc outside this line; neutral hydrogen (Wright et al., 1972) follows closely the HII regions (the Westerbork observers have other examples of the close correlations HI-HII in various galaxies). This segregation is predicted by the density-wave theory of spiral structure (see the lecture by F. Shu and for example Shu et al., 1972): according to this, stars form along the edge of the arm which may (not in all cases) be a large-scale shock, seen as a dust lane because of the strong compression of gas and dust. If the wave rotates more slowly than the matter (this is the case for M33), it leaves behind it the evolving stars and clusters. It is obvious that such effects have to be looked for in external galaxies and in our galaxy as tests for the density-wave theory; there is, however, surprisingly little known in this field.

With all these effects in mind, we can look at the problem of determining the spiral structure close to the sun, by looking at the distribution of spiral tracers. A good discussion is given by Bok (1971). The most interesting method of all uses HII regions and their exciting stars: we can link directly in this case with a fair accuracy distances obtained by photometry of the exciting stars with kinematics (radial velocity of the gas). A lot of work in this direction has been done at the Marseille observatory (see e.g. Georgelin and Georgelin, 1971). Fig. 1, taken from Moffat and Vogt (1973), provides a picture of the spiral structure in

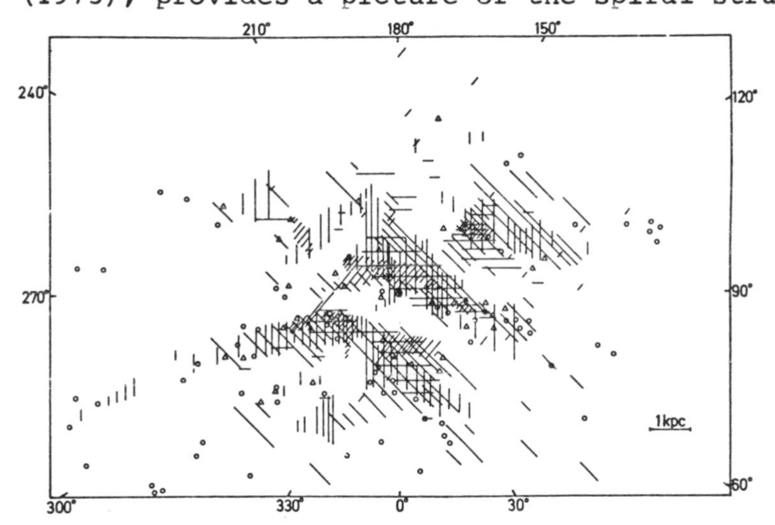

Fig. 1: Galactic structure based on different spiral arm tracers. The hatched regions correspond to tracers with sufficient frequency to allow a determination of an approximate density distribution according to the following code: Hatching direction 0° (|)=HII regions after Georgelin and Georgelin (1971). Hatching direction 45° (\\)=supergiants after Humphreys (1970). Hatching direction 90° (-) =emission stars of types Bpe and B0/0.5 (III-V)e after Schmidt-Kaler (1964). Hatching direction 135° (/)=young open star clusters. Individually marked are: o=Wolf rayet stars after Smith (1968). ▲ = δ-Cep stars with periods

longer than 15 days after Tammann (1970). The sun is indicated by an asterisk. (After Moffat and Vogt, 1973)

the vicinity of the sun where the distributions of all studied spiral tracers have been superimposed; most of this work can be found in the IAU Symposium No. 38. Most of it relates to stars or HII regions; relatively little is known for HI and dust because of the lack of distance criteria. Three arms are visible; one (the local arm, also called the Orion or Cygnus arm) contains the sun which is located near the inside of the arm. At 2-3 kpc in the direction of the galactic center we find the Sagittarius arm and in the opposite direction the Perseus arm. How these arms are related to each other is far from clear; they are quite patchy and contain elongated concentrations which are not always even elongated along what one would like to be an arm.

As we said, relatively little is known of the local distribution of neutral hydrogen. There is certainly a gross agreement between the distribution of neutral hydrogen and the picture of Fig. 1, but the details are very uncertain. The only features which do not depend on kinematics are the concentrations of HI in opposite directions in Cygnus ($l=80^{o}$) and in Vela ($l \simeq 265^{o}$): one gets the feeling that one is looking along the local arm. Weaver (1970) has given radio evidence that the local arm in the direction of Cygnus merges finally with the Sagittarius arm. There are several recent studies of local neutral hydrogen, resting mainly upon observations of gas at high galactic latitudes, which must be mainly local. For example, Fejes and Wesselius (1973) conclude that part of this gas is located in a disk tilted by about 45° with respect to the galactic plane, with a radius 100-200 pc, and a mean density 0.4 cm^{-3}; this thin (35 pc) structure is immersed in a thicker, more plane-parallel distribution of hydrogen. Lindblad et al. (1973) discuss a feature very extended in the sky, which may be an expanding shell associated with the Gould Belt, of about 160 and 330 pc semi axes in the galactic plane, the center of which is 140 pc distant in the direction $l^{II}=150^{o}$. Thus, the local situation looks complicated.

An interesting point is the existence of "anomalies" in the kinematics of HI. For example, the 21-cm Perseus arm seems, if we use Schmidt's rotation curve, further away than the "optical" Perseus arm, and looks less inclined and more circular (Roberts, 1972): this is probably not real and indicates non-circular motions; furthermore, optical interstellar velocities in front of O stars of the Perseus arm show anomalies correlated with the 21-cm velocities, suggesting that some gas is departing from circular motion and approaches along the line-of-sight with vebcities up to 20 km/s (Fig. 2). Agreement is good with an arm rotating at $\Omega_p \simeq 12.5$ km $s^{-1}kpc^{-1}$ with a pitch angle of 8 to 12°. The high-spatial resolution study of the 21-cm absorption profile of Cassiopeia A

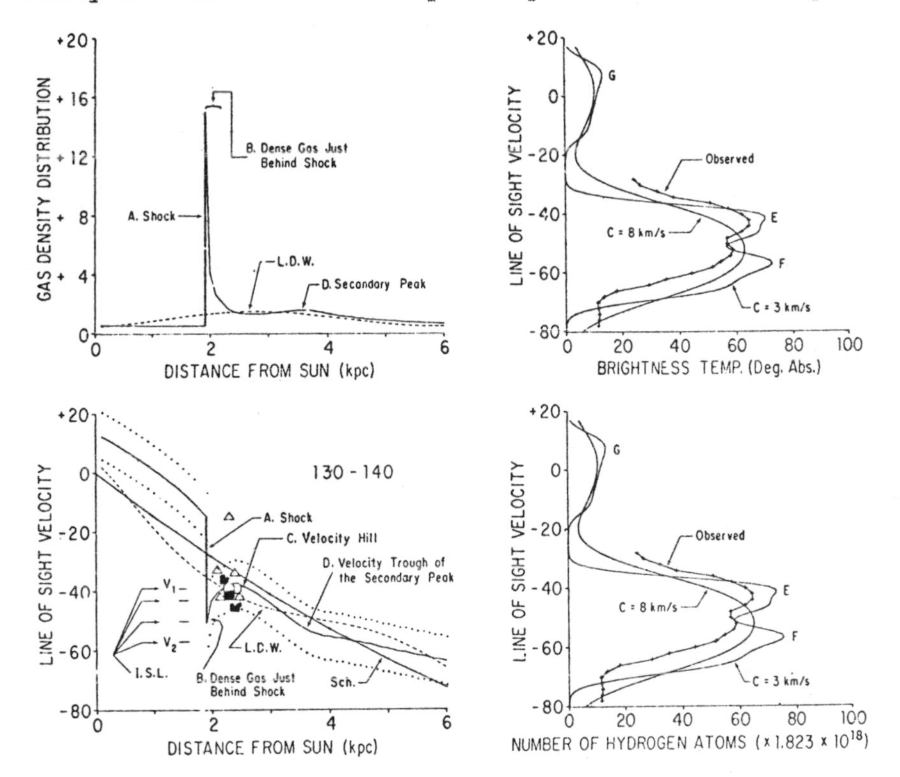

Fig. 2: Theoretical features of the 12° TASS model in the longitude range $l=130^o-140^o$. Lower left, line-of-sight velocity versus distance from the sun. The basic motion of the gas is from left to right. Sch (solid curve with crosses), equilibrium Schmidt model. L.D.W. (-----), linear density-wave model. Heavy solid curve (———), TASS model. The two dotted curves (......) outline a dispersion band of ± 8 km s^{-1} about the TASS curve; O, O associations; △, O-B2 open clusters; ■, HII regions. Having evolved into their luminous stages of evolution some 5-50 million years after the initial stages of their triggering at the shock (Region B), these young luminous objects have migrated with the gas to Region C, several hundred parsecs beyond the shock. The interstellar absorption-line data

of Münch (ISL) provide the strongest observational evidence for systematic motion along the Perseus arm. Here the galactic shock appears as a rather essential feature in order to account for these lines, particularly those at large negative velocities (e.g. V_2). Upper left, gas density distribution $\sigma = (\sigma_o + \sigma_1)/\sigma_o$, versus distance from the sun. Lower right, upper right, theoretical number-density and brightness-temperature profiles, respectively. The profiles for mean turbulent dispersion speed c=3 km s^{-1} and c=8 km s^{-1} are plotted along with the observed brightness-temperature profile (solid curve with crosses) derived from Westerhout (1966, 1969). Splitting of the overall HI feature in the profile into multiple subcomponents and peaks is a rather natural tendency in the presence of a shock. In the profile for c=3 km s^{-1}, the contribution to peak E comes almost entirely from the effect of "velocity crowding" which occurs in Region C where a substantial amount of rarefied gas is located along a large distance over the velocity hill. Here the effect of the galactic velocity field on the profile is so important that it cannot be underemphasized. On the other hand, the contribution to peak F comes from both Region B, the dense gas just behind the shock, and Region D, the less dense gas in the secondary peak. (After Roberts, 1972)

---

by E. Greisen (1973) seems also to indicate the presence of very small and dense clouds in the Perseus arm at the expected post-shock velocity. However, a recent discussion by Verschuur (1973a) questions the whole picture.

The local arm may also present evidence for a shock wave: there is a cold (20 K) cloud seen in 21-cm absorption in front of nearly all objects in the direction of Sagittarius between $l = 345^o$ and 25°, with a velocity of + 7 km/s (Quirk and Crutcher, 1973). This cloud also shows up by interstellar absorption lines in front of close-by stars like ζ Oph and is at less than 150 pc; there is also a concentration of dust in this region (Uranova, 1970). This feature is interpreted by Quirk and Crutcher as dense gas behind a galactic shock, which would have a tilt angle of 5-8° (very different from the tilt angle of 30° for the optical local arm; the reason of the discrepancy is unclear).

So far, the preceding observations are the only ones close to the sun which can be taken as proofs of the density-wave theory (see also Minn and Greenberg, 1973). Local velocity gradients could also produce detectable effects, like a rotation of the nodes of the sin 2l variation of the radial velocity (Rohlfs, 1972); however, this effect is not clearly established by observation (see Takakubo, 1967; Mebold, 1972; Falgarone and Lequeux, 1973).

The complicated and badly known structure of the interstellar gas close to the sun should be kept in mind when interpreting the results of the next paragraph.

2. Density and distribution of neutral hydrogen

Clouds and the intercloud medium. Clark (1965) suspected first from his observations of the 21-cm line in emission and in absorption that the interstellar medium might consist in relatively dense and cold clouds immersed in a hotter and more tenuous intercloud medium. This has received ample confirmation through two major studies of compared 21-cm profiles in emission and in absorption (Hughes et al., 1971; Radhakrishnan et al., 1972), and to a lesser degree through examinations of high-latitude 21-cm surveys. The clouds give sharp (2-3 km/s) emission profiles, and somewhat sharper absorption profiles when observed in front of distant radio sources. From the brightness temperature and the optical depth, temperature and column densities can be derived: typical values are 60-80 K and 3×10^{20} atom cm^{-2}, respectively, but there is a very large scatter around these values; temperatures between 10 K and more than 100 K have been measured and column densities are between 3×10^{19} (a lower observational limit) and 10^{21} atom cm^{-2} (indeed the densest clouds consist mainly of molecules). Because of the lack of distance criteria, the diameters of the clouds are virtually unknown, and the notion of a "standard" cloud of dimensions 10 pc, etc., does not make much sense. There certainly exists a large variety of clouds, including small "cloudlets", large complexes with masses $\gg 10^3$ $M_\odot$, and even sheets (see e.g. Van Woerden, 1967).

In nearly all directions at medium and high latitudes the intercloud medium gives broad ($\gtrsim$ 18 km/s) and weak (1-20 K) 21-cm emission profiles. Recent unpublished observations by Lazareff and the author show that this medium is sometimes seen in weak absorption, sometimes not, with spin temperatures ranging from a few hundred degrees K to more than 1-2000 K: it may be that intermediates between a very hot intercloud medium and cold clouds exist as suggested by Pottasch (1972), in agreement with the time-dependent theory of heating-ionization of the interstellar medium (see Dalgarno and McCray, 1972).

Thus, it seems that the distinction clouds-intercloud medium may be somewhat arbitrary; in what follows, we use the definition of Radhakrishnan et al. (1972): medium not visible in absorption in their observations, with broad emission profiles. There are a few attempts to determine the temperature of the intercloud medium (so defined). From the width of the diffuse Hα and Hβ lines observed in various directions, Reynolds et al. (1973) find $T \leqslant 6000$ K. Hachenberg and Mebold (1972), and Baker (1973) using a statistical study of intercloud profiles in extended fields, find $T \simeq 5000$ K; there are rather large uncertainties in this method, although it looks promising.

Up to now, we have little idea from observation whether the interarm medium differs from the intercloud medium which we see close to the sun at high galactic latitudes. Quirk (unpublished) estimates $\langle n_H \rangle \simeq 0.2$-$0.3$ cm^{-3} and $T > 200$ K in the interarm medium, which, if this is true, is not significantly different from the local intercloud gas.

<u>Mean density and z-distribution of clouds and the intercloud medium</u>. From their observations, Radhakrishnan et al. (1972) estimate for the clouds $N_H/T \simeq 1.5 \times 10^{19}$ atom cm^{-2} K per kiloparsec in the galactic plane, $\langle T \rangle \simeq 60$ K and $N_H \simeq 9 \times 10^{20}$ atom cm^{-2} kpc^{-1}, corresponding to a mean density in the plane $n_H = 0.29$ cm^{-3}. Since the mean column density of clouds perpendicular to the plane is about 1.5×10^{20} atom cm^{-2}, the equivalent thickness of the cloud system at the sun is about 1/3 kpc; but we do not know the shape of their z-distribution. From observations of the latitude profile of distant 21-cm arms and from theory, this is presumably close to gaussian (Kellman, 1972). Hughes et al. (1971) arrive partially to the same conclusions. Radhakrishnan et al. (1972) also state that there are 3 clouds per kpc with τ (21 cm) $\geqslant 0.5$. Observers of optical interstellar lines come out with many more clouds per kpc, but the contradiction is only apparent. Radioastronomers observe clouds with column densities $3 \times 10^{19} < N_H < 10^{21}$ atom cm^{-2}, giving visual absorptions $0.02 < A_V < 0.6$ magnitudes but the weakest of these clouds already give strong H and K or D interstellar lines. The velocity dispersion between individual clouds has been discussed by many authors, most recently Falgarone and Lequeux (1973), and is of the order of 6.4 km s^{-1} (1-dimensional). There are some evidences that the velocities of the clouds follow an exponential law (Mast and Goldstein, 1970).

The determination of the thickness of the intercloud medium is a more complicated story and must rest on the differential rotation; we refer to the papers by Mebold (1972) and Falgarone and Lequeux (1973). The results, from the latter paper, are given in Table 1, which compares the properties of the intercloud medium and of the clouds. The density and thickness of the intercloud medium assume that its z-distribution is gaussian, a somewhat controversial statement.

	clouds	intercloud medium
density at z=0	0.29 cm^{-3}	0.16 cm^{-3}
equivalent thickness	330 pc	580 pc
column density at $b=90^{\circ}$	1.5×10^{20} cm^{-2}	1.4×10^{20} cm^{-2}
mean temperature	60 - 80 K	several hundred to 6000 K, perhaps 5000 K ?
velocity dispersion	6.4 km s^{-1}	$\lesssim$ 7 km s^{-1} ?

Table 1: Parameters of the distribution of clouds and the intercloud medium close to the sun.

It should be noticed that there is now a good agreement between the total mean density of local neutral gas (0.45 atom cm^{-3}) derived from radioastronomy and the average density derived from Ly α absorption profiles obtained by the OAO 2 satellite (Savage and Jenkins, 1972). Most of the apparent earlier disagreement was due to an incorrect determination of the Ly α equivalent widths, and partly to invalid comparison of 21-cm and Ly α data.

Another interesting comparison is with the very local density of neutral gas as derived from Ly α scattering in the solar system ("interstellar wind"). Although there are large errors in the derived parameters, these observations suggest $n_H \simeq 0.1-0.2$ cm^{-3}, $T \simeq 10^3-10^4$ K (Bertaux et al., 1972): thus, the sun is rather immersed in the intercloud medium.

3. Density and distribution of the ionized gas

HII regions. Their distribution is visible in Fig.1. It is likely that only a very minor fraction of the ionized gas is in "classical" HII regions, most of it being distributed in the intercloud medium, where its mass amounts to 15% of the total mass of the gas in the vicinity of the sun.

Free electrons in the general medium. That the general interstellar matter is ionized is shown by a number of radio and optical observations (see Dalgarno and McCray, 1972). The best value of the mean electron density is provided by the dispersion of pulsars for which distances are known independently, mostly through 21-cm absorption (see Guélin, 1973; Falgarone and Lequeux, 1973). These observations give $\langle n_e \rangle \simeq 0.03\ cm^{-3}$ with relatively little scatter around this value, which refers to distances 1 to 2 kpc from the sun, in the galactic plane, where most of the pulsars are located. n_e is probably barely larger in clouds, although this point is somewhat controversial (see Ball et al., 1972, the lectures by Pottasch; Pottasch, 1972; and Reynolds et al., 1973). Most of the other observational quantities refer to $\langle n_e^2 \rangle$ rather than $\langle n_e \rangle$ with always some weighting by a power of the electron temperature, and are thus much less easy to handle. However, the study of optical recombination lines from the diffuse medium (Reynolds et al., 1973) suggests values of $\langle n_e^2 \rangle^{1/2}$ of the order of 0.05-0.1, barely larger than $\langle n_e \rangle$ and thus indicating relatively little clumping of the local diffuse ionized matter (this is not necessarily true in the central regions of the galaxy; it does not imply, on the other hand, that the neutral gas is not clumpy: in current theories of ionization by cosmic rays or soft X-rays, the electron density is almost independent of the atomic density).

Determining the thickness of the electronic disk is a somewhat lengthy affair to discuss here, and the reader is referred to Falgarone and Lequeux (1973). The result of their study is that although the thickness of the electron disk is poorly determined, it is of the order of 1 kpc or more and thus much thicker than the neutral gas z-distribution. The column density of electrons seen from the sun at $b=90^o$ is about $4 \times 10^{19}\ cm^{-3}$, thus the total mass of ionized gas is about 1/6 of the total mass of neutral gas close to the sun. The ionization degree of the intercloud medium varies from 15% at z=0 to large values at large z. This is consistent (in a steady ionization model) with an ionization rate of a few times $10^{-15}\ s^{-1}$ per hydrogen atom. Reynolds et al. (1973) arrive at a similar number from the intensity of the diffuse H β line, a method which is almost model-independent.

4. The galactic magnetic field

There are several well-known evidences for the existence of a galactic magnetic field: polarization of the stellar light and of the non-thermal radio continuum, Zeeman effect in the 21-cm line, Faraday rotation, as well as more indirect evidences (synchrotron radiation, influence on supernova remnants, Razin effect on the low-frequency galactic radiation, etc.).

Optical interstellar polarization has been observed for a large number of stars (Mathewson and Ford, 1970; Klare and Neckel, 1970) and exhibits a complex behavior. The distribution of the polarization vectors in the sky is rather irregular, with extended regions with high polarization far from the galactic plane, but this is limited to relatively near-by stars (distance < 400 pc). For larger distances, the vectors are more parallel to the galactic plane, but for some longitudes where they show a tangled structure suggesting that one is looking along a tube of force. Mathewson (1968) interprets the local structure as a flattened helical magnetic field, immersed in a more regular field which dominates at distances $\simeq$ 500 pc from the sun. However, this interpretation assumes that the spurs of non-thermal radio radiation, which correspond to the highly polarized regions at high galactic latitudes, are parts of this helical structure. However, there are increasing evidences (Berkhuijsen, 1973) that the spurs are old supernova remnants, and Spoelstra (1973) has even attempted to derive the direction of the (more or less uniform) magnetic field from the study of its interaction with the spurs considered as supernova remnants.

At larger distances from the sun the magnetic field is obviously more uniform. Its direction can be derived from observations of the run with galactic longitude of interstellar polarization, Faraday rotation of pulsars (Manchester, 1972 and 1973) and quasars (Reinhardt, 1972; Mitton and Reinhardt, 1972; Falgarone and Lequeux, 1973), but with discrepant results which are summarized in Table 2.

The often quoted statement that the magnetic field lies along spiral arms thus cannot be considered well established.

The best determination of the intensity of the local magnetic field probably comes from the study of pulsars for which both the Faraday rotation measure RM $\sim \int n_e \; B_{\parallel} \; dl$

Method	Longitude of field direction	radius around the sun
Optical polarization	80° (+ local helical field ?)	300 to 4000 pc
Spurs and loops	40°	$\leqslant$ 300 pc
Faraday rotation of pulsars	90° to 110°	300 pc (mean)
Faraday rotation of quasars	110°	$\leqslant$ 2 - 3 kpc
Local arm (gas)	60° - 80°	$\sim$ 1 kpc
Local arm (stars)	50°	$\sim$ 1 kpc

Table 2

and the dispersion measure $DM \sim \int n_e \, dl$ has been determined (Fig. 3). The ratio RM/DM gives for each pulsar the mean value of $B_{//}$ along the line of sight, weighted by n_e (since n_e is not too different in clouds and the inter-cloud medium, this is actually the mean magnetic field that is measured). With the assumption of a uniform field, one finds by a statistics on pulsars its direction (quoted above) and mean intensity: Manchester (1972) gives $\langle B \rangle \simeq 3.5$ microgauss and Reinhardt (1972) $\langle B \rangle \simeq 2.2$ microgauss. These are lower limits since the field lines of force may be tangled, and the true B is probably between 3 and 5 microgauss. Recent determinations of the local synchrotron radio emissivity of the galaxy, 3-10 x 10^{-41} W Hz^{-1} m^{-3} at 400 MHz (see e.g. Baldwin, 1967; Alexander et al., 1970; Brown, 1973), when combined with the known local flux of cosmic ray electrons which are responsible for this emission, also provide $B \simeq 5$ microgauss.

An important point is the weakness of the magnetic field in clouds as measured by Zeeman effect on the 21-cm line (Verschuur, 1970, 1971). Positive results have been obtained in a few directions and give fields of 10-70 microgauss in clouds with estimated densities 100-1000 atoms cm^{-3}, and negative results ($B \lesssim 5$ to 10 microgauss) in many other clouds. When compared with the average $B \simeq 5$ microgauss corresponding to a density $\simeq 0.4$ atom cm^{-3}, it is clear that the law of variation $B \sim n_H^{2/3}$ expected if the clouds collapse isotropically from the general interstellar medium and are perfectly conducting, does not hold, contrary to the claim of Verschuur (1970); possible explanations are a) the field

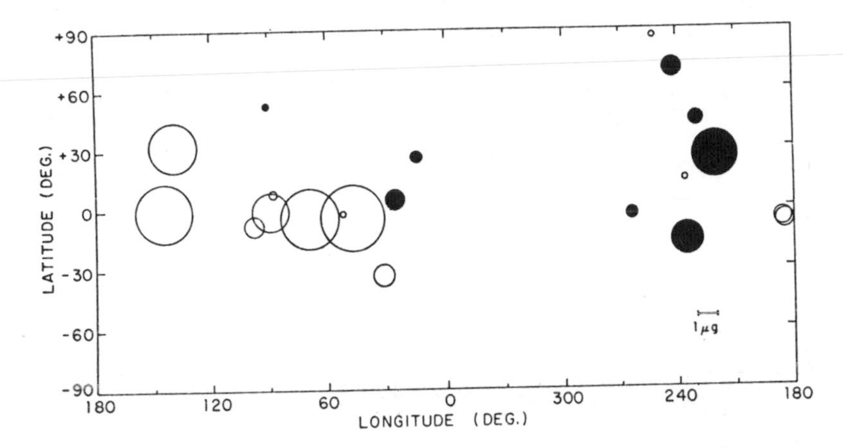

Fig. 3: Mean line-of-sight magnetic field components for pulsars plotted in galactic coordinates. For fields greater than 0.3 microgauss the circle diameter is proportional to the field strength; for positive rotation measures (field toward the observer) the circles are filled, whereas for negative rotation measures (field away from the observer) they are open. The diameter for a 1-microgauss field is indicated in the figure. (After Manchester, 1972)

is much tangled in the clouds and the measured quantity, $\langle B_{\parallel} \rangle$, is much smaller than B; b) the clouds do not contract isotropically, but preferentially along the lines of force, thus taking rather the form of sheets.

II. GENERAL DISTRIBUTION OF INTERSTELLAR MATTER

1. Spiral structure and kinematics

Spiral structure and the arm-interarm contrast. At large distances from the sun, distances can be estimated only from radial velocities (mostly from 21-cm and radio recombination lines; OH, H_2CO and CO lines are also increasingly used for spiral structure). Since one expects from any spiral arm theory, in particular the density-wave theory, to observe deviations from circular velocities, distances and thus the spiral structure are difficult to be obtained separately from the knowledge of kinematics. The old isodensity maps of neutral hydrogen distribution, obtained on the assumption of pure rotation, are now very hard to believe and all that we are presently able to obtain is a rough sketch of the spiral structure. Even the two main recent pictures of the spiral structure obtained by Kerr at Maryland and Weaver at Berkeley, although very sketchy, are different in several respects (see Simonson, 1970) (Fig. 4). It is now clear that the problems of space distribution and of

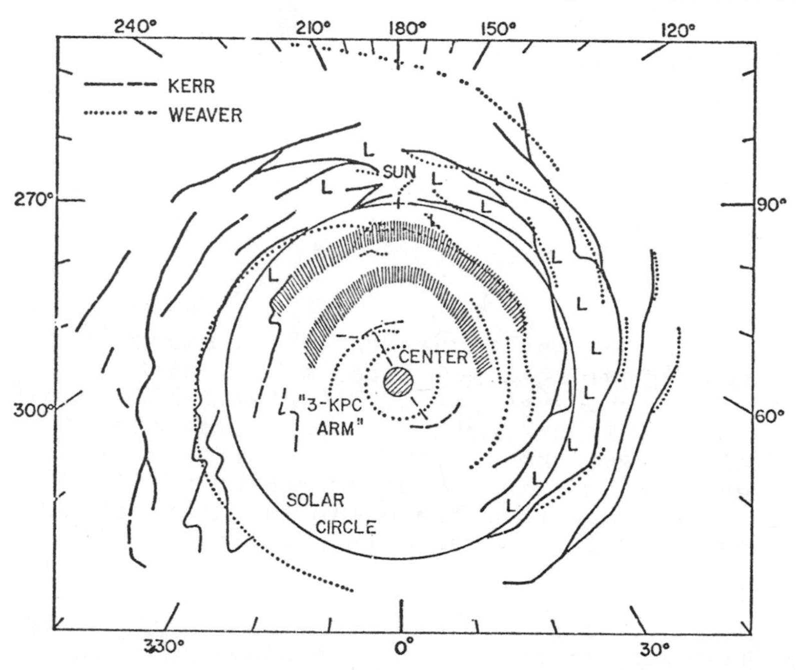

Fig. 4: Superimposed maps of the neutral hydrogen distribution in the galaxy as interpreted by Kerr (1969) and by Weaver (1970). The points marked "L" denote regions that Kerr interprets as markedly deficient in neutral hydrogen. Dashed lines and hatched areas indicate regions where the location is uncertain, not regions where the hydrogen is weak or strong. The radius of the solar circle is 10 kpc. Galactic longitude is shown aronnd the edge of the map. (After Simonson, 1970)

kinematics must be attacked simultaneously. The most general line of approach has been taken by Burton (1972b), who has explored in detail the influence of the velocity field on the 21-cm profiles. He has tried to fit actually observed 21-cm profiles in several directions in the range $l = 90^{\circ}$ to 120°, first with a model with pure rotation plus random motions allowing only for density variations, and secondly with the other extreme model with uniform density and deviations from circular motion. Both models give a fair fit (Fig. 5), but the first one

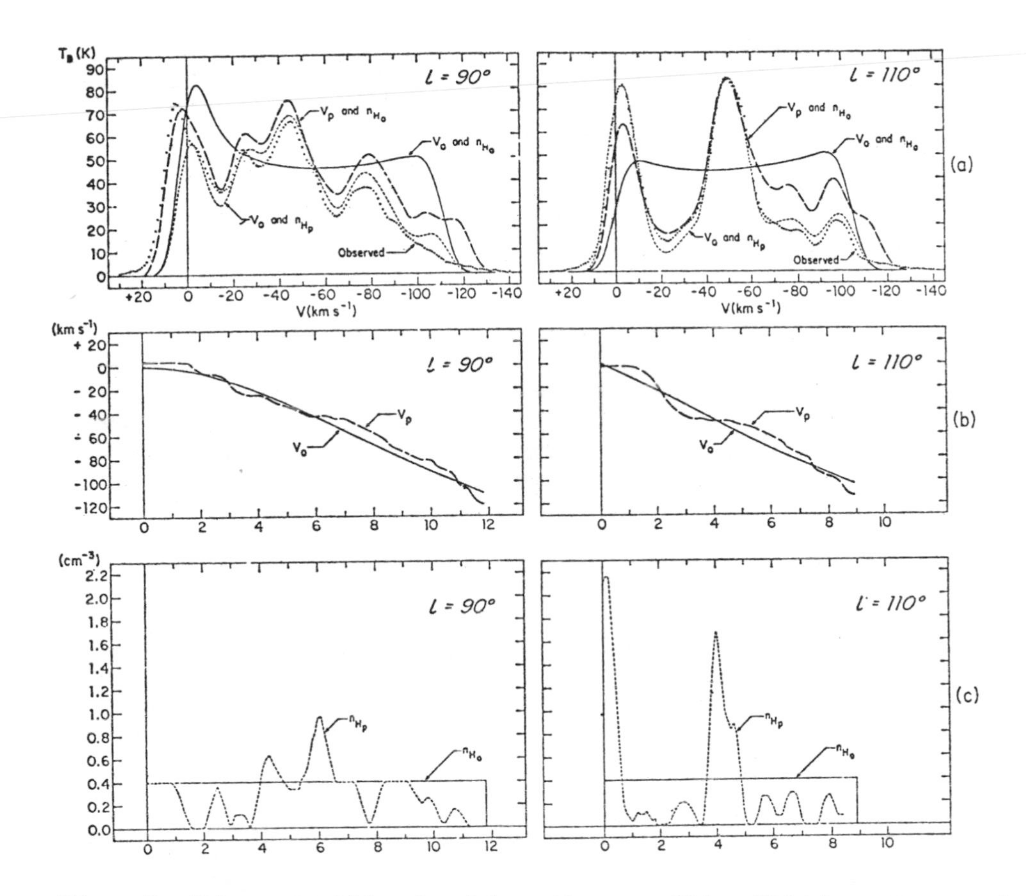

Fig. 5: Diagram illustrating the profile-fitting approach described in the text for two typical longitudes. In the 4 lower figures, the abscissa is r in kpc. a Comparison of the observed profiles (dots) with model profiles. The zero-order profile (full-drawn line) was calculated with the basic circular-rotation velocity field $V_o(r)$ and a completely uniform hydrogen distribution n_{Ho}=0.4 cm^{-3}. The heavy dashed-line profile was calculated by perturbing only the velocity field, giving $V_p(r)$, leaving the density uniform. The profile illustrated by the light dashed line was fit to the observations by varying the density, giving $n_{Hp}(r)$, while retaining the circular velocity field, $V_o(r)$. b Perturbed line-of-sight velocity-field, $V_p(r)$, which together with a uniform hydrogen distribution results in the heavy dashed-line profile in a. Note that the kinematic approach can account for the "forbidden-velocity" peak at l=90° in a natural manner and that the required streaming amplitudes appear reason-

able. c Perturbed density distribution, $n_{Hp}(r)$, which together with the circular-rotation velocity field results in the light dashed-line profile in a. Note that the density approach fails to account for the "forbidden-velocity" hydrogen and that extended, essentially empty, interarm regions are necessary. (After Burton, 1972b)

---

requires densities going to zero in the interarm region, an unrealistic result, and is unable to reproduce the "forbidden" part of the spectrum at negative velocities. It is likely that the truth is somewhere between these extreme models. A step further is to fit a spiral-structure model to the observations, as done by Burton in several papers (Burton and Shane, 1970; Burton, 1972a and b; Yuan, 1969; etc.). The density wave theory gives a good fit to the observations (yielding $\Omega p=12.5$ km s^{-1} and a pitch angle of 15° for the Perseus arm between 90° and 120°; in this self-consistent model the arm/interarm density contrast is 7/1). But this <u>alone</u> cannot be taken as a <u>proof</u> of the validity of the density-wave theory, and the arm-interarm contrast is one of the most badly known parameters of the gas distribution. It is probable that a better check of the theory and better determinations of the arm/interarm contrast will come from high-resolution 21-cm and optical studies of external galaxies.

Another kind of complication comes from the non-homogeneous nature of the interstellar medium. Contrary to what is usually assumed when studying galactic structure, the clouds have a wide range of temperatures and the optical depth corrections are most uncertain. Moreover, the intercloud medium is optically thin and it is conceivable that in some directions its brightness temperature will be larger than the temperature of clouds and that the latter will appear in absorption in front of the intercloud medium (spiral arms could then appear as minima in the profiles rather than peaks!). Such an extreme situation probably does fortunately not often occur in practice.

<u>Tracing of spiral arms</u>. The two recent models of spiral structure have been derived on the assumption of pure rotation by Kerr (1969) and Weaver (1970), respectively. Fig. 5, from Simonson (1970), shows both sketches superimposed. Their differences are not as large as it would appear at first glance and are mainly related to the interpretation of the tangential points of the spiral structure and to the connection between

them; the discrepancies are discussed in Bok (1971) and Simonson (1970). The most interesting, because the corresponding features are best used for determining the pitch angle of the galactic structure, is relative to the connection or absence of the connection of the Sagittarius arm and the Carina feature (which is seen tangentially at $l=283^o$). There is a gap in the neutral hydrogen between $l=292^o$ and 305°, that Kerr takes as an evidence against this connection, while Weaver attributes it merely to an irregularity in a single Sagittarius arm. An abundant literature discusses this point: see in particular Georgelin and Georgelin (1970). Weaver's model gives a pitch angle for the Sagittarius-Carina feature of 9°, but a mean pitch angle on the galaxy of 12° to 15°, in agreement with Burton's (1972b) determination in longitudes 90° - 120°; Kerr's model has more circular arms, with a pitch angle of perhaps 7°.

Optical tracers of the spiral structure are increasingly used in relatively transparent regions like Carina, and in the region $l=300^o$ to 330°, where one can see up to 6 kpc from the sun (Chu-Kit, 1973).

A large effort has been made to obtain radial velocities of a large number of distant HII regions using their radio recombination lines (Mezger, 1970). Because of the weakness of the lines, only "giant" HII regions have been studied: they lie mainly between 4 and 13 kpc from the center of the galaxy (plus some very close to the center). However, no clear spiral picture emerges from these observations, mostly because the distance ambiguity between the two positions symmetrical with respect to the tangential points, which exist for kinematic distances in the inner regions of the galaxy, has not been solved for a large fraction of these HII regions.

The longitude distribution of HII regions (seen by their continuum radio emission), and of the non-thermal galactic background have also been considered for delineating the tangential points of the inner spiral arms. However, the results obtained in this way are not very convincing.

2. Radial distribution of the interstellar gas

This is illustrated in Fig. 6, from Mezger (1972), for neutral hydrogen and giant HII regions. The distribution of HII regions in general, as derived from the

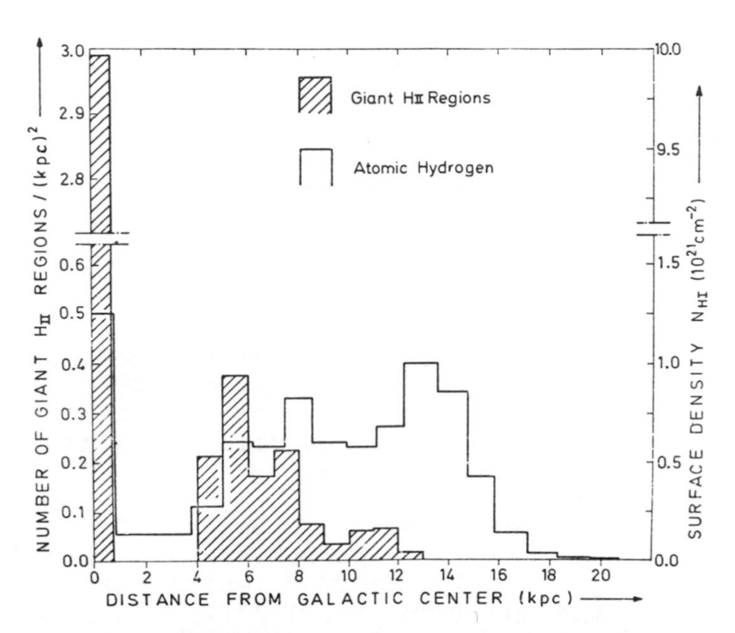

Fig. 6: Distribution of neutral hydrogen and of giant HII regions as a function of distance from the galactic center. Both diagrams give surface densities obtained by integration over the coordinate z perpendicular to the galactic plane. (After Mezger, 1972)

longitude distribution of the thermal radio background, is very similar to that of giant HII regions. It is seen that neutral hydrogen extends very far from the center, and shows a hole in the central regions, as in many external galaxies. Between 3 and 8 kpc from the center, recent observations of radio recombination lines in the general medium show the presence of a large distributed amount of ionized hydrogen (see e.g. Gordon and Cato, 1972; Gordon et al., 1972; Davies, to be published). However, it is far from clear whether these lines come from ionized gas in a more or less uniform intercloud medium, in small HII regions or even in cold clouds (see e.g. Cesarsky and Cesarsky, 1973). Observations of free-free absorption in front of galactic radio sources in this direction also show the existence of a large quantity of ionized gas within 8 kpc from the center, but do not solve the problem (Dulk and Slee, 1972).

Supernova remnants and the non-thermal radio emission show approximately the same radial distribution as giant HII regions (Ilovaisky and Lequeux, 1972), but the z-extent of the non-thermal emission is very large.

3. z-distribution of interstellar gas

The z-distribution of atomic hydrogen is illustrated in Fig. 7 from a combination of various sources: it is seen that the thickness of the galactic disk increases very much with increasing radius. This is probably a consequence of the variation with galactic radius of the galactic gravitational potential (Kellman, 1972).

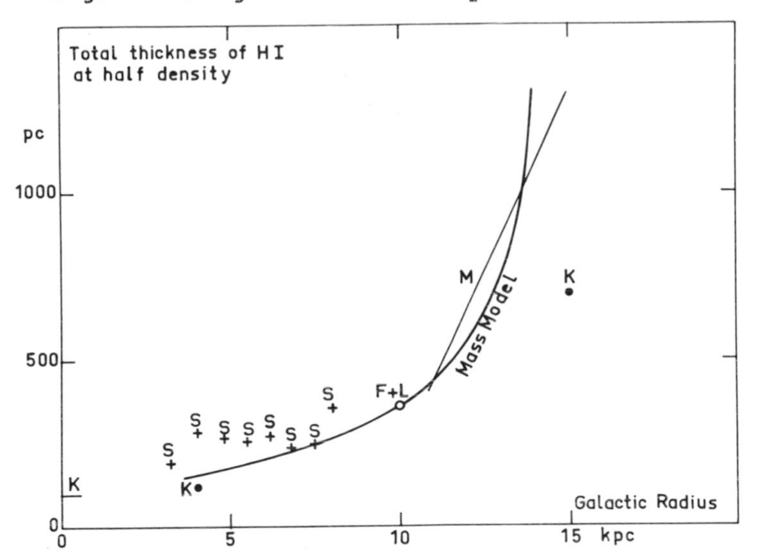

Fig. 7: Thickness of the disk of neutral hydrogen as a function of distance from the galactic center. Letters represent determinations from various references: K=Kerr (1969); S=Schmidt,M., 1957: Bull.Astron.Inst.Netherl.13, 247, after correction for the new galactic distance scale; M=McGee,R.X., Milton, J.A., 1964: Austr.J.Phys. 17, 128; F + L=Falgarone and Lequeux (1973). The curved line is the run of thickness with galactic radius as predicted by Kellman (1972) from the Innanen's galactic mass model.

Extreme population I objects show the same phenomenon, but they make an even thinner layer (Ilovaisky and Lequeux, 1972), probably because the rate of stellar formation is proportional to a power of the gas density larger than unity.

At large distances from the galactic center, there are considerable distortions of the galactic 21-cm disk and one finds neutral hydrogen far from the galactic plane (Kerr, 1969; Oort, 1970). These deviations do not occur closer to the center than the sun where they are at most 20 pc (Varsavsky and Quiroga, 1970). Some of these asymmetrical high-latitude extensions correspond to extensions of known arms (Kepner, 1970; Wannier et al., 1972). Other features are interpreted by Verschuur (1973b) and Davies (1972) as corresponding to extensions of even more distant arms. These distortions and extensions are generally interpreted as due to tidal interaction of the Magellanic clouds, which, if this is true, should have passed recently (about 5 x 10^8 years ago) at less than 20 kpc from the galactic center (Hunter and Toomre, 1969; Eneev et al., 1972). Similar distortions and also large departures from circular velocities have been observed in the external parts of M33, where they could have been produced by the interaction of M31 (Huchtmeier, 1973), and in other galaxies.

Verschuur (1973b) and Davies (1972) have presented very similar models and give an interpretation of the intermediate and high velocity clouds observed at medium and high glactic latitudes (Oort, 1966, 1967) as gas torn from the galaxy (by interaction of the Magellanic clouds) and falling back on the galactic disk. This is a rather convincing picture, but a crucial test would be to measure the distance of these clouds from the sun, a very difficult task, however.

III. THE GALACTIC CENTER REGION

We define more or less arbitrarily this region as extending up to the expanding feature at 4 kpc from the center. This is a very peculiar region in many respects. Most of neutral gas is in the form of huge molecular clouds, while there is little atomic hydrogen. Ionized gas is limited to the very central regions where it forms giant HII regions with somewhat particular properties. Dust is very abundant in the central region of the galaxy. The kinematics of the whole region is complicated and often dominated by non-circular motions, indicating some sort of strong activity in the galactic center.

1. Neutral gas

Neutral atomic hydrogen. Its distribution and kinematics have been extensively studied in the literature with the following results (see e.g. Kerr, 1967).

A disk-like structure extends to about R = 800 pc from the center and has a maximum rotation velocity of 230 kpc s^{-1} and no apparent expansion. The half-density thickness is 100 pc in the inner parts and 250 pc at the edge. The H density is approximately uniform and equal to 0.3 at cm^{-3}, the total mass is about 4 x 10^6 $M_\odot$. Outside the disk, the structure is complex and badly known, with violent non-circular motions. At about R = 4 kpc we encounter a somewhat complex ring (or arm) expanding with a velocity of about 50 km s^{-1}. Its thickness is about 120 pc and its total mass is estimated as 2x10^7 $M_\odot$ (very uncertain), as much as the total mass of neutral hydrogen inside 4 kpc but only 0.5% of the total mass of galactic interstellar matter.

There are large quantities of gas with peculiar velocities above and below the central disk; the gas at positive longitudes is mostly below the plane, and the gas at negative longitudes above it. From the work of van der Kruit (1970, 1971), Sanders et al. (1972), Sanders and Wrixon (1972), it appears that there are several complexes of clouds more or less symmetrical with respect to the center, with different orientations, masses and velocities. The most important complex has a mass of about 10^6 $M_\odot$ and velocities of 130 km s^{-1}, if interpreted as an ejection from the galactic nucleus; their direction is at small angle from the galactic plane. Another complex has a mass of 10^5 $M_\odot$ and is at about 50° from the plane; it coincides with features in the continuum radiation noticed by Kerr (1967). Whether this gas is continually pushed from the center at low velocities or is ejected by a central explosion at larger velocities is unclear. Eventually, it falls back on the galactic plane (Kerr, 1967), and causes expansion motions in the plane; the expanding feature at 4 kpc may be explained in this way (van der Kruit, 1971), but it may also simply correspond to the inner Lindblad resonance ring in the density-wave model of spiral structure. Table 3 summarizes the energies and time-scales appropriate to the different observed ejections.

The rate of kinetic energy production is rather moderate: for comparison, a supernova explosion produces about 10^{51} ergs and the rate 6 x 10^{42} erg s^{-1} would correspond to the energy of 1 supernova every 10 years if the energy is produced in supernovae.

Molecules. The gas in the galactic center region is mostly molecular. The distribution of OH (McGee, 1970), H_2CO (Scoville et al., 1972) and more recently CO (Solo-

Clouds	Present Mass M⊙	Initial kinetic energy (ergs) if explosive	Age (years)	mean rate of energy production (ergs s^{-1})
HI: Van der Kruit (1970)	10^6	4×10^{53}	6×10^6	$3 \times 10^{39-40}$
HI: Sanders - Wrixon (1970)	10^5	4×10^{52}	4×10^5	
4 kpc arm	2×10^7	10^{55}	10^6	3×10^{42}
Molecular clouds	10^8?	2×10^{55}?	10^6	6×10^{42}?

Table 3

mon et al., 1972; Scoville and Solomon, 1973) is now relatively well known. The distributions of all three molecules look rather similar, and it is clear that they are mostly concentrated in large (30 pc), dense (> $10^3 H_2$ molecules cm^{-3}) and massive (10^4-10^6 M$_\odot$ or more) clouds (Scoville and Solomon, 1973; Fomalont and Weliachew, 1973). A lower limit to the total mass can be given by noticing that the CO lines appear strongly only when the density is more than 10^3 H_2 molecules cm^{-3}, and is 10^8 M$_\odot$: this is two orders of magnitude more than the mass of neutral hydrogen in the same region. Molecular clouds are close to the center, within a radius 200 - 300 pc, and show radial motions with velocities 70 to 145 km/s; the kinetic energy of a hypothetical explosion which would have given this phenomenon is indicated in Table 3. There is some suggestion that these clouds may form a ring (Scoville, 1972). The thickness of the molecular system is badly known, probably about 60 pc.

2. Ionized gas

The situation with the ionized gas has been summarized by Lequeux (1967) and Mezger et al. (1972). The ionized gas is contained in an extended source extending over $1^\circ \times 0^\circ 4$ (170 x 70 pc), with a mass less than 7×10^5 M$_\odot$ (this region emits broad recombination lines and is probably responsible for the free-free absorption in front of the central radio source Sgr A, which has a cut-off at 100 MHz: Dulk, 1970). There is also a number of giant HII regions, mostly at positive longitudes, which are certainly close to the center (plus a few non-thermal radio sources). The kinematics of this hydrogen has been studied by Mezger et al. (1972), and is consistent with pure rotation of a disk or ring with a radius 150 pc at most, but the rotation is not consistent with the predictions of the mass distribution of Sanders and Lowinger (1972). The physical state of this gas causes problems; it is very highly turbulent, as shown by the width of the recombination line. There may be a deficiency of helium with respect to hydrogen, as the helium recombination lines have never been detected in the HII regions (Churchwell and Mezger, 1973; Huchtmeier and Batchelor, 1973). On the other hand, there is some suggestion from X-ray absorption data that oxygen may be overabundant, but this is poorly established (Seward et al., 1972).

3. Dust and infrared emission

The galactic nucleus is heavily obscured by 27 magnitudes of visual absorption (Becklin and Neugebauer, 1968), but the central stellar cluster begins to show up at 2.2 μ . Amongst the complex structures seen in the infrared, several may be attributed to dust emission: a 1 pc diameter strong source at 10-20 μ (but here the mechanism for heating is not known) (Becklin and Neugebauer, 1969), several discrete sources at 100 μ (coinciding with HII regions and/or molecular clouds), and finally an extended source at 100 μ of dimensions $3^{\circ}6 \times 2^{\circ}$ (600 x 350 pc) (Low and Aumann, 1970; Hoffmann et al., 1971; Gezari et al., 1973). The total mass of dust giving these emissions is difficult to estimate because its temperature and nature are not really known; however, the dust-to-gas ratio is not necessarily higher than in the rest of the galaxy. A puzzling fact is that the extended infrared source is quite a lot more extended in galactic latutude than the gas.

Fig. 8 summarizes the complex distribution of the different components of the interstellar gas and dust in the galactic center region. Some information on the distribution of non-thermal radio emission is also given. A discussion of the origin of these structures is outside the scope of this review.

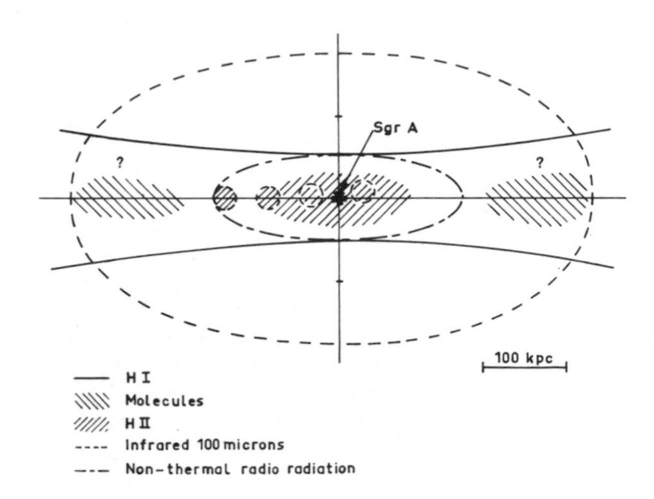

Fig. 8: Structure of the central part of our galaxy.

REFERENCES

Alexander, J.K., Brown, L.W., Clark, T.A., Astron.Astrophys. 6, 476, 1970.

Baker, P.L., Astron.Astrophys. 23, 81, 1973.

Baldwin, J.E., IAU Symposium No 31, 337, 1967.

Ball, J.A., Cesarsky, D., Dupree, A.K., Goldberg, L., Lilley, A.E., Astrophys.J.Lett. 162, L25, 1970.

Becker, W., Fenkart, R.B., IAU Symposium No. 38, 205, 1970.

Becklin, E.E., Neugebauer, G., Astrophys.J. 151, 145, 1968.

Becklin, E.E., Neugebauer, G., Astrophys.J.Lett. 157, L31, 1969.

Berkhuijsen, E., Astron.Astrophys. 24, 143, 1973.

Bertaux, J.L., Ammar, A., Blamont, J.E., Space Research XII, 1559, 1972.

Bok, B.J., Highlights of Astronomy 2, 73, 1971.

Brown, L.W., Astrophys.J. 180, 359, 1973.

Burton, W.B., Astron.Astrophys. 16, 158, 1972a.

Burton, W.B., Astron.Astrophys. 19, 51, 1972b.

Burton, W.B., Shane, W.W., IAU Symposium No. 38,397,1970.

Cesarsky, C.J., Cesarsky, D., preprint, 1973.

Chu-Kit, M., Astron.Astrophys. 22, 69, 1973.

Churchwell, E., Mezger, P.G., Nature, 242, 319, 1973.

Clark, B.C.G., Astrophys.J., 142, 1398, 1965.

Courtès, G., Dubout-Crillon, R., Astron.Astrophys. 11, 468, 1971.

Dalgarno, A., McCray, R.A., Ann.Rev.Astron.Astrophys. 10, 375, 1972.

Davies, R.D., Mon.Not.R.Astron.Soc. 160, 381, 1972.

Dulk, G.A., Astrophys.Lett. 7, 137, 1970.

Dulk, G.A., Slee, O.B., Austral.J.Phys. 25, 429, 1972.

Eneev, T.M., Kozlov, N.N., Sunyaev, R.A., Astron.Astrophys. 22, 41, 1973.

Falgarone, E., Lequeux, J., Astron.Astrophys., in press, 1973.

Fejes, I., Wesselius, P., Astron.Astrophys. 24, 1, 1973.

Fomalont, E.B., Weliachew, L., Astrophys.J., in press, 1973.

Georgelin, Y.P., Georgelin, Y.M., Astron.Astrophys. 7, 133, 1970.

Georgelin, Y.P., Georgelin, Y.M., Astron.Astrophys. 12, 482, 1971.

Gezari, D.Y., Joyce, R.R., Simon, M., Astrophys.J.Lett. 179, L67, 1973.

Gordon, M.A., Brown, R.L., Gottesman, S.T., Astrophys. J. 178, 119, 1972.

Gordon, M.A., Cato, T., Astrophys.J. 176, 587, 1972.

Greisen, E.W., Astrophys.J., in press, 1973.

Guélin, M., Proc.I.E.E.E., in press, 1973.

Hachenberg, O., Mebold, U., Communication at the First European Conference of the IAU, Athens, 1972.

Hoffmann, W.F., Frederick, C.L., Emery, R.J., Astrophys. J.Lett. 164, L23, 1971.

Huchtmeier, W., Astron.Astrophys. 22, 91, 1973.

Huchtmeier, W., Batchelor, R.A., Nature 243, 155, 1973.

Hughes, M.P., Thompson, A.R., Colvin, R.S., Astrophys. J.Suppl. 23, 323, 1971.

Hunter, C., Toomree, A., Astrophys.J. 155, 747, 1969.

Ilovaisky, S.A., Lequeux, J., Astron.Astrophys. 18, 169, 1972.

Kellman, S.A., Astrophys.J. 175, 353, 1972.

Kepner, M., Astron.Astrophys. 5, 444, 1970.

Kerr, F.J., IAU Symposium No. 31, 239, 1967.

Kerr, F.J., Annual Rev.Astron.Astrophys. 7, 39, 1969.

Klare, G., Neckel, T., IAU Symposium No. 38, 449, 1970.

Kruit, P.C. van der, Astron.Astrophys. 4, 462, 1970.

Kruit, P.C. van der, Astron.Astrophys. 13, 405, 1971.

Lequeux, J., IAU Symposium No. 31, 393, 1967.

Lindblad, P.O., Grape, K., Sandqvist, A., Schober, J., Astron.Astrophys. 24, 2, 1973.

Low, F.J., Aumann, H.H., Astrophys.J.Lett. 162, L79, 1970.

Lynds, B.T., IAU Symposium No. 38, 26, 1970.

Manchester, R.N., Astrophys.J. 172, 43, 1972.

Manchester, R.N., Bull.A.A.S. 5, 35, 1973.

Mast, J.W., Goldstein, S.J.,Jr., Astrophys.J. 159, 319, 1970.

Mathewson, D.S., Astrophys.J.Lett. 153, L47, 1968.

Mathewson, D.S., Ford, V.L., Memoirs Roy.Astron.Soc. 74, 139, 1970.

McGee, R.X., Austral.J.Phys. 23, 541, 1970.

Mebold, U., Astron.Astrophys. 19, 13, 1972.

Mezger, P.G., IAU Symposium No. 38, 107, 1970.

Mezger, P.G., Course on Interstellar Matter, Saas-Fee, 1972.

Mezger, P.G., Churchwell, E.B., Pauls, T.A., Communication at the First European Conference of the IAU, Athens, 1972.

Minn, Y.K., Greenberg, J.M., Astron.Astrophys. 24, 393, 1973.

Mitton, S., Reinhardt, M., Astron.Astrophys. 20, 337,1972.

Moffat, A.F.J., Vogt, N., Astron.Astrophys. 23, 317, 1973.

Oort, J.H., Bull.Astron.Inst.Netherl. 18, 421, 1966.

Oort, J.H., IAU Symposium No. 31, 279, 1967.

Oort, J.H., IAU Symposium No. 38, 142, 1970.

Pottasch, S.R., Astron.Astrophys. 20, 245, 1972.

Quirk, W.J., Crutcher, R.M., Astrophys.J. 181, 359, 1973.

Radhakrishnan, V., and associates, Astrophys.J.Suppl. 24, 1972.

Reinhardt, M., Astron.Astrophys. 19, 104, 1972.

Reynolds, R., Roesler, F., Scherb, F., Astrophys.J. 179, 651, 1973. See also Bull.A.A.S. 5, 24.

Reynolds, R., Scherb, F., Roesler, F., Astrophys.J.Lett. 181, L79, 1973.

Roberts, W.W.,Jr., Astrophys.J. 173, 259, 1972.

Rohlfs, K., Astron.Astrophys. 17, 246, 1972.

Sanders, R.H., Lowinger, Th., Astron.J. 77, 292, 1972.

Sanders, R.H., Wrixon, G.T., Astron.Astrophys. 18,467,1972.

Sanders, R.H., Wrixon, G.T., Penzias, A.A., Astron.Astrophys. 16, 322, 1972.

Savage, B.D., Jenkins, E.B., Astrophys.J. 172, 491, 1972.

Scoville, N.S., Astrophys.J.Lett. 175, L127, 1972.

Scoville, N.S., Solomon, P.M., Thaddeus, P., Astrophys. J. 172, 335, 1972.

Scoville, N.S., Solomon, P.N., Astrophys.J. 180, 55, 1973.

Seward, F.D., Burginyon, G.A., Grader, R.J., Hill, R.W., Palmieri, T.M., Astrophys.J. 178, 131, 1972.

Shu, F.H., Milione, V., Gebel, W., Yuan, C., Goldsmith, D.W., Roberts, W.W., Astrophys.J. 173, 35, 1972.

Simonson, S.C.,III, Astron.Astrophys. 9, 163, 1970.

Solomon, P.M., Scoville, N.S., Jefferts, K.B., Penzias, A.A., Wilson, R.W., Astrophys.J. 178, 125, 1972.

Spoelstra, T.A.Th., Astron.Astrophys. 24, 149, 1973.

Takakubo, K., Bull.Astron.Inst.Netherl. 19, 125, 1967.

Uranova, T.A., IAU Symposium No. 38, 228, 1970.

Van Woerden, H., IAU Symposium No. 31, 3, 1967.

Varsavsky, C.M., Quiroga, R.J., IAU Symposium No. 38, 147, 1970.

Verschuur, G.L., IAU Symposium No. 38, 150, 1970.

Verschuur, G.L., Astrophys.J. 165, 651, 1971.

Verschuur, G.L., Astron.Astrophys. 24, 193, 1973a.

Verschuur, G.L., Astron.Astrophys. 22, 139, 1973b.

Wannier, P., Wrixon, G.T., Wilson, R.W., Astron.Astrophys. 18, 224, 1972.

Weaver, H., IAU Symposium No. 38, 126, 1970.

Wright, M.C.H., Warner, P.J., Baldwin, J.E., Mon.Not. R.Astron.Soc. 155, 337, 1972.

Yuan, C., Astrophys.J. 158, 871, 1969.

LIST OF REFERENCES FROM WHICH THE FIGURES HAVE BEEN TAKEN

Fig. 1: Moffat and Vogt, Astronomy and Astrophysics 23, p. 317; figure 2 p. 319, 1973.

Fig. 2: Roberts, The Astrophysical Journal, 173, p. 259: figure 3 p. 268, 1972.

Fig. 3: Manchester, The Astrophysical Journal, 172, p.43: figure 2 p. 51, 1972.

Fig. 4: Simonson, Astronomy and Astrophysics 9, p. 163: figure 1 p. 166, 1970.

Fig. 5: Burton, Astronomy and Astrophysics 19, p. 51: figures 2 a b c p. 54, 1972.

Fig. 6: Mezger, unpublished course in Saas Fee, 1972.

Figures 7 and 8 are original.

DENSITY-WAVE THEORY OF SPIRAL STRUCTURE

F.H. Shu+

State University of New York at Stony Brook
New York, USA

I. INTRODUCTION

The topics I would like to cover in this lecture range from the problem of spiral structure on a large scale to the problem of star formation on a small scale. I shall adopt the density-wave interpretation of these diverse phenomena as the unifying point of view. This point of view originated with B. Lindblad and has been developed into a coherent form by C.C. Lin and his associates (see the review by Lin, 1971).

As is well-known, a wave-interpretation of spiral structure in disk galaxies is required because of the so-called "persistence dilemma". If the spiral arms of a differentially rotating galaxy were always composed of the same material, the arms would wind up into progressively tighter trailing spirals. An additional turn would be introduced about every 10^8 years; such overwinding can be statistically ruled out on observational grounds. The only viable alternative seems to be to identify the spiral arms with the crest of a density wave.

II. STELLAR DYNAMICAL THEORY

Since stars constitute the bulk of the mass in the disk of a galaxy and since the large-scale coherence of

---

+Alfred P. Sloan Foundation Fellow, 1972-74

the spiral structure is attributed in the density-wave theory to gravitational forces, the dynamics of the disk stars play a central role. Thus, I shall first review the salient results of the theory for the stellar disk.

The theory of stellar density-waves has been developed primarily in the linear, i.e. small-amplitude, regime because the excess mass density associated with the crest of the density-wave amounts typically to only a small fraction of the equilibrium mass density of the stars. However, some aspects of the nonlinear perturbations of stellar orbits near resonant conditions have been studied by Contopoulos (1970) and by Vandervoort (1973).

A. Global modes of "warm" stellar disks

Kalnajs (1965, 1971) and Shu (1968, 1970a) have considered the normal modes of oscillation of a completely flattened stellar disk containing finite dispersions of the velocity components in the plane of the disk. They showed that the linear problem can be formulated in terms of a homogeneous integral equation (cf. eq.(32) of Shu, 1970a):

$$\tilde{\omega} S(\tilde{\omega}) = \int_0^\infty K_{m\omega}(\tilde{\omega}, a)\, a S(a)\, da \tag{1}$$

in the above equation, $S(\tilde{\omega})$ is the radial part of the perturbation surface density, $\sigma_1(\tilde{\omega},\theta,t) = S(\tilde{\omega}) e^{i(\omega t - m\theta)}$. The kernel $K_{m\omega}(\tilde{\omega},a)$ is a definite functional of the equilibrium phase-space distribution of the stars, but its explicit form need not concern us here.

Attempts have been and are being made to solve the integral equation (1), or some simplification of it, to find modes possibly exhibiting global instability. Realistic models of the equilibrium disk require numerical techniques, but Kalnajs (1972; see also Hunter, 1963) has discovered analytical solutions for a singular class of equilibrium models whose basic mass distribution corresponds to that of a uniformly rotating disk. This study reinforces results found in n-body calculations which indicate that velocity dispersions considerably in excess of that required to cure axisymmetric instabilities (Toomre, 1964) may be needed to stabilize the barlike instability (Miller et al., 1970; Hohl, 1970; Peebles and Ostriker, 1973; see also the study of Ostriker and

Bodenheimer, 1973 on the stability of rotating gaseous masses).

However, before one accepts the conclusion that a large amount of velocity dispersion must be uniformly present in at least some subpopulation of the stellar component of the galaxy, we believe two issues need to be clarified. (1) To what extent does a large concentration of mass in the central regions help to stabilize the tendency of the disk to form oval distortions? After all, as extreme examples, neither the configuration of the planets in the solar system nor Saturn's rings seem to be in any danger of assuming a barlike shape. (2) To what extent does the possible development of strong galactic shocks in the gaseous component of the galaxy limit the unstable barlike mode? In particular, can a quasi-stationary state eventually be reached which could reasonably represent the state of affairs in a barred-spiral galaxy?

B. Properties of stellar density-waves in the WKBJ approximation

Numerical techniques have not, so far, proven very successful for the study of spiral structues which are characteristically found in normal spiral galaxies. Here, the WKBJ approximation introduced by Lin and Shu (1964) still provides the most powerful method of analysis for the solution of equation (1).

In the lowest order of approximation for tightly-wrapped spiral waves, Lin and Shu (1966) obtained - for a stellar disk of infinitesimal thickness and relatively small velocity dispersion - a relation which gives the local radial wavenumber k once the wave frequency ω and angular symmetry m are known:

$$D(\omega,k,m,\tilde{\omega}) = \frac{|k|}{k_T} \frac{F_\nu(x)}{(1-\nu^2)} = 1 \qquad (2)$$

In the above equation, $\nu = (\omega-m\Omega)/\varkappa$ is the wave frequency measured by a particle traveling with circular frequency $\Omega(\tilde{\omega})$ in units of the epicyclic frequency $\varkappa(\tilde{\omega})$; $k_T = \varkappa^2/2\pi G\sigma_{*0}$ is the characteristic wavenumber of a rotating disk of basic surface density $\sigma_{*0}(\tilde{\omega})$ (Toomre, 1964); and $x = k^2\langle c_{\tilde{\omega}}^2\rangle/\varkappa^2$ is a dimensionless measure of the mean-square dispersive speed of the stars in the radial direction. The "reduction factor" $F_\nu(x)$ has the functional form

$$F_\nu(x) = \frac{(1-\nu^2)}{x}\left[1 - \frac{\nu\pi}{\sin\nu\pi}\;\frac{1}{2\pi}\int_{-\pi}^{\pi}\cos(\nu s)e^{-x(1+\cos s)}ds\right] \quad (3)$$

In the next order of the WKBJ approximation, Shu (1970b) found that the amplitude of the radial component of the spiral gravitational field, kA, is distributed in accordance with the relation

$$\tilde{\omega} A^2(\tilde{\omega}) \frac{\partial \ln D}{\partial \ln k} = \text{constant} \quad (4)$$

if we assume, for simplicity, that ω is purely real.

Shu (1968) and Vandervoort (1970) considered the modifying effects of finite disk thickness. They found that the dispersion relation (2) could be generalized to

$$D(\omega,k,m,\tilde{\omega}) = \frac{|k|}{k_T}\;\frac{F_\nu(x)}{(1-\nu^2)}\;T = 1 \quad (5)$$

where T is the "reduction factor due to finite disk thickness". In the "crude approximation" of Vandervoort, T has the functional form

$$T = (1+|k|z_o)^{-1} \quad (6)$$

where z_o is the effective half-thickness of the stellar disk.

Shu (unpublished) has also shown that an extension of Dewar's (1972) variational-principle derivation of equation (4) - valid for a disk of infinitesimal thickness - yields a formally identical result for a disk of finite thickness. The only differences are that one must substitute the expression (5) for D and interpret k^2A^2 as the mass-weighted average over the thickness of the disk of the square of the radial component of the spiral gravitational field. Equations (4) and (5) allow, then, rather realistic theoretical calculations of spiral patterns once we are given an equilibrium model of the stellar disk and an empirical estimate of the pattern speed, $\Omega_p = \omega/m$ (see, e.g. Lin et al., 1969; Shu et al., 1971).

C. Propagation and absorption of spiral density-waves

Toomre (1969) has shown that a group of spiral waves of the type considered by Lin and Shu would propagate in the radial direction with a group velocity which would lead to their eventual disappearance in about 10^9 years. In particular, short trailing waves in the galaxy would propagate inwardly until they reached the inner Lindblad resonance ring (located between 3 and 4 kpc). Mark's (1971) linear analysis shows that they would be absorbed there by a plasma-like process of resonance damping. Thus, the outstanding dynamical problem at the present time is to account for the origin and permanence of the spiral density-wave.

Before reaching the Lindblad resonances, the propagation of a group of spiral waves occurs with the conservation of "wave action" (see Toomre's 1969 interpretation of the amplitude relation found by Shu). Kalnajs (private communication) has discovered that this conservation principle is equivalent to the conservation of wave energy and angular momentum. Dewar (1972) has demonstrated that the conservation principle can also be given an alternative derivation by applying variational methods to the "averaged Lagrangian" as discussed for general WKBJ-wave systems by Whitham (1965).

In the notation introduced previously, the conservation principle, for the time-dependent problem, reads

$$\frac{\partial E}{\partial t} + \frac{1}{\tilde{\omega}} \frac{\partial}{\partial \tilde{\omega}} (\tilde{\omega} c_{g\tilde{\omega}} E) = 0 \tag{7}$$

In equation (7) E and $c_{g\tilde{\omega}}$ are, respectively, the angle-averaged energy-density and the radial component of the group velocity of the wave:

$$E = \frac{|k| A^2}{8\pi G} \omega \frac{\partial D}{\partial \omega} , \quad c_{g\tilde{\omega}} = \frac{\partial D/\partial k}{\partial D/\partial \omega} \tag{8}$$

Clearly, equation (7) reduces to equation (4) for steady waves.

In the WKBJ approximation, the conservation principle (8) includes the conservation of angular momentum since the relation between the densities of wave energy E and angular momentum J reads

$$J = \frac{m}{\omega} E = \Omega_p^{-1} E \tag{9}$$

The density of wave energy in a frame which rotates with the material, i.e. which rotates with angular velocity $\Omega(\tilde{\omega})$, is given by the expression

$$H = E - \Omega J \tag{10}$$

and is positive-definite. It can be estimated to have the approximate numerical value of 1×10^9 erg cm^{-2} as an average at the solar circle. This corresponds to an energy density which is only comparable to the projected energy densities contained locally in turbulent gas motions, starlight, cosmic ray particles, and the interstellar magnetic field. Thus, the ordering tendency of the spiral gravitational field can be expected to dominate in the interior of the galaxy where its strength is relatively larger, but we should not be surprised if the local spiral structure is somewhat messy.

D. The possible origin of spiral structure

From equations (9) and (10) and the knowledge that H is positive-definite, we easily show that E and J are negative inside of corrotation, and positive, outside. (The corrotation circle occurs where $\Omega(\tilde{\omega}) = \Omega_p$; we believe this to happen at about 15 kpc in our own galaxy.) This suggests that a transfer of positive energy and angular momentum outward across the corrotation circle (or equivalently, a transfer of negative energy and angular momentum inward) could, in principle, provide a possible mechanism for the spontaneous growth of the wave amplitude everywhere. The sense of the transport of energy and angular momentum by short trailing spiral waves is in the correct sense for this process. Further study is needed before we will know whether such transport plays an essential role in the generation of spiral structure in normal spiral galaxies (cf. Lynden-Bell and Kalnajs, 1972).

In any case, it may be premature to concentrate at the present time on any one specific mechanism to generate or to maintain spiral density waves. We should also keep in mind the feedback mechanism of Lin (1970) which involves a wavelike oval distortion of the central regions (see also Feldman and Lin, 1973); external forcing by orbiting companions (Toomre and Toomre, 1972); the role of resonant interactions (Lynden-Bell and Kalnajs, 1972); and the ejection of massive gas clouds from the nucleus of a spiral galaxy (Van der Kruit et al., 1972).

Admitting that the problem of the origin of spiral structure remains largely unsolved, we still find it profitable to proceed to work out some of the further implications of the density-wave theory. Foremost in importance for this Study Institute is the picture which has emerged concerning the present-day formation of stars in spiral galaxies. This is the topic of the second half of my lecture.

III. GALACTIC SHOCKS AND STAR FORMATION

When we examine carefully a photograph of a spiral galaxy such as M 81 or NGC 5364, a number of important questions come to mind. How do stars know to form along twin spiral arms which are very narrow in width and yet extend many kpc in length? Why should the brilliant spiral arms of the newly-born OB stars be accompanied at a small distance apart by the dark lanes?

In the density-wave theory, the fundamental spatial coherence of the spiral structure is provided by an orderly spiral gravitational field. This field is supported primarily by the disk stars; it amounts typically to some several percent to that of the basic axisymmetric field. Thus, the first question raised above can be rephrased: How can the zones of active star formation in spiral galaxies be defined so clearly by a spiral gravitational field of such small amplitude?

The results obtained for the analytical linear theory of the material response contain a hint. These imply that the surface density response of a particular component of the galaxy is roughly proportional to the inverse square of the characteristic speed. Since the effective acoustic speed of the interstellar gas is typically one-third to one-fourth of the r.m.s. random velocity of the disk stars, the <u>same</u> spiral field which induces only a small fractional variation of the disk stars could lead to a very large density response in the interstellar gas.

However, no simple extrapolation of the linear theory adequately explains why the lanes of high gas compression should be so <u>narrow</u> as to lead to HII regions arranged amazingly like "beads on a string" in galaxies like M 81 or NGC 5364. Roberts (1969; see also Fujimoto, 1966) showed that this puzzle can be resolved by considering the nonlinear theory of gaseous density-waves - in particular by allowing for the possibility of <u>galactic shocks</u>. The concept of galactic shocks has received subsequent observational support from the work of Mathewson et al. (1971) on the pattern of synchrotron emission from M 51.

A. Nonlinear gaseous density-waves and galactic shocks

In dealing with the large-scale pattern of gas flow in a galaxy, we can adopt a coarse model in which we smooth out the clumpy nature of the interstellar medium and treat it as a homogeneous gas with an effective speed of sound of about 8 km s^{-1}. If we further adopt the asymptotic approximation that the variation of the perturbation quantities is much larger in the direction normal to spiral equipotentials than along them, we find that the nonlinear theory for the gaseous response to forcing by a linear stellar density-wave is characterized by three dimensionless parameters. These are: $\nu = (\omega - m\Omega)/\varkappa$, the intrinsic wave frequency; $\alpha = (k^2 + m^2\tilde{\varpi}^{-2})^{1/2} a/\varkappa$, the intrinsic acoustic speed; and $f = (k^2\tilde{\varpi}^2 + m^2) A/\tilde{\varpi}^2\varkappa^2$, a scaled field strength (cf. definitions of symbols after eq. (2)). For a given pattern speed, all three parameters may vary with radial position in the galaxy. (Note, however, that the linear theory of stellar density-waves specifies only the relative variation of f and not its absolute magnitude; see eq. (4). Shu et al. (1973) have shown that variation of the three independent parameters ν, α, and f lead to the following three distinct phenomena.

1. Natural development of galactic shocks. If we consider the flow properties along a given circle (ν and α held constant) and increase f slowly, the steady-state numerical solutions evolve continuously from the linear solutions given by Lin et al. (1969), through the forced analogues of the slightly nonlinear solutions of Vandervoort (1971), to the highly nonlinear shocked solutions of Roberts (1969). Thus, galactic shocks arise as a necessary consequence of the theory for waves of sufficiently large amplitude; and the shockwave solutions obtained by Roberts and others are simply the highly nonlinear counterparts of the usual linear density-waves. The actual development in time of galactic shocks has been followed by Woodward (1973).

2. Narrow versus broad arms. For $|\nu| > \alpha$, which corresponds to the case when the phase-velocity of the stellar density-wave measured with respect to the unperturbed circular motion is greater than the effective acoustic speed of the interstellar gas, we recover Roberts' result that the regions of high gas compression are very narrow in the directions perpendicular to spiral equi-

potentials. If star formation is related to high gas density, the lanes where HII regions would be found would be very narrow since the typical lifetime of an HII region, about 10^7 years, is much shorter than the time it takes to traverse from one peak of gas compression to another, on the order of 10^8 years. Moreover, because of the delay in star formation, the lanes of HII regions would appear displaced downstream from the regions of highest gas compression. The latter are marked by the lanes of dark matter. The predicted displacement of HII regions from the dark lanes is in accordance with B.T. Lynds' (1970) observations.

For the previously unexamined regime $|\nu| < \alpha$, the regions of high gas compression are broad. In our theory, then, the distinction between broad optical arms and narrow optical arms depends on whether the Doppler-shifted phase-velocity of the stellar density-wave is less than or greater than the effective acoustic speed of the interstellar gas. Much of the body of a given galaxy may be characterized by one situation or the other. To be specific, our experience leads us to expect that broad arms - together with open spiral structures - should prevail in galaxies with weak differential rotation, whereas narrow arms - together with tightly-wound spiral structures - should prevail in galaxies with strong differential rotation.

3. Secondary spiral features. For a range of radii where $\nu^2-\alpha^2 = n^{-2}$ ($n = \pm 2, \pm 3, ...$) is approximately satisfied, the nonlinear response calculations show that secondary compressions of the interstellar gas can be produced. This is a distinctly nonlinear resonance effect and is to be contrasted with the Lindblad resonances which correspond to $n = \pm 1$. To be precise, if we use the terminology of the people who work in nonlinear oscillations and who think in terms of periods rather than frequencies, the nonlinear resonances should be called "ultraharmonic resonances". We have suggested that the $n^{-1} = -1/2$ resonance, the most prominent nonlinear resonance, may explain the nature of the Carina spiral feature in our own galaxy, whereas the higher-order ultraharmonic resonances may explain the spurs and feathers often seen in the outer regions of external galaxies.

B. Relative rates of star formation in the galaxy

By adopting a picture where the present-day formation

of stars occurs as a result of the compression in a galactic shockwave, we are able to explain certain observational features which would otherwise remain quite inexplicable. A primary example of this is illustrated in Fig. 1. This figure presents the well-known discrepancy in the Milky Way system between the abundance distributions of HII gas and HI gas. The HII/HI ratio looks as plotted because Mezger found the number of giant HII regions to be strongly peaked at 4 kpc while Van Woerden found the HI abundance distribution to be fairly flat from 4-15 kpc (see Mezger, 1970).

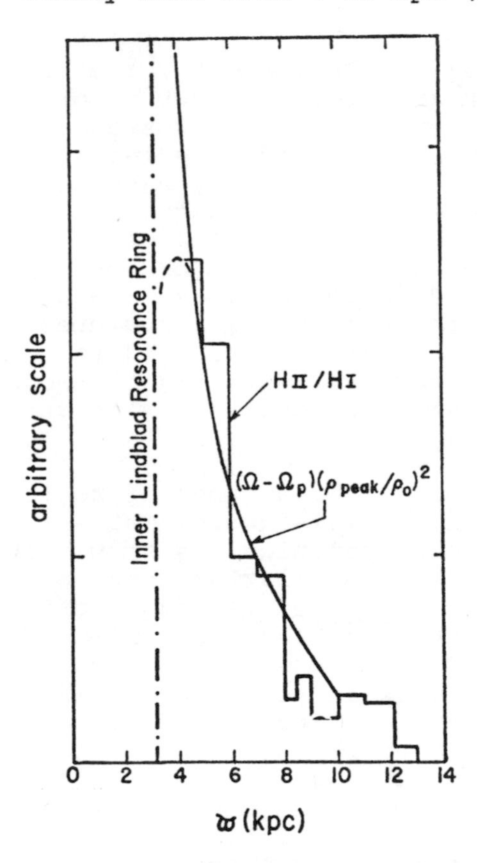

Fig. 1: The abundance distribution of HII relative to that of HI in the galaxy. The histogram gives the observational ratio as derived from the measurements of Van Woerden and Mezger. The theoretical curve is based on calculations of the nonlinear gaseous response.

The discrepancy between the abundance distribution of HII and HI has been cited to indicate that there exists a factor other than average gas density which controls the present-day formation of stars in the galaxy. The density-wave theory provides a natural explanation for this factor.

The rapid increase inward HII/HI until 4 kpc can be attributed theoretically to two sources (Shu, 1973): (a) the kinematical increase inward the frequency $(\Omega-\Omega_p)$ at which the interstellar gas is periodically compressed, and (b) the dynamical increase inward the effective peak compression (P_{peak}/P_o) of the interstellar gas. The latter is partly associated with the increase inward the strength of the spiral gravitational field measured as a fraction of the basic axisymmetric field in accordance with equation (4), and partly with the highly nonlinear amplification calculated for the peak density response of the interstellar gas. Finally, there is little star formation inside 3-4 kpc because stellar density waves cannot propagate inside the inner Lindblad resonance ring. (The conditions right at the galactic center are presumably special.)

C. Compression of the interstellar medium on a small scale

The description of star formation given above is phenomenological; it does not provide a mechanistic explanation of the actual process. To do this requires a more detailed model of the interstellar medium. In particular, we must recognize explicitly the clumpy nature of the actual distribution of interstellar gas. In our work, we adopted the two-phase model of the interstellar medium (Shu et al., 1972).

Upon compression, the intercloud gas in the two-phase models of Pikelner (1967), Field et al. (1969), and Spitzer and Scott (1969) would remain nearly isothermal with the temperature staying near 10^4 °K. Hence, the calculations based on a gas with an isothermal speed of sound of 8 km s^{-1} give, as direct calculations have verified, a fairly good representation of the behavior of the intercloud gas - apart from the possibility of phase transitions. However, previous work represents the galactic shock as a discontinuity in the flow whose end states are related through "isothermal jump conditions". The actual shock layer has, of course, finite width and structure because the gas is thrown badly out of equilibrium inside the shock layer, and finite time is required for the gas to relax to another equilibrium consistent with the downstream flow conditions. This structure can be calculated in some simple cases.

1. Structure of a galactic shock layer. For illustrative purposes, Fig. 2 shows the structure of the shock layer in the intercloud medium when F, the spiral field strength measured as a fraction of the axisymmetric field, equals 3.3% at 10 kpc from the galactic center. In Fig. 2 we have ignored interactions with the cloud

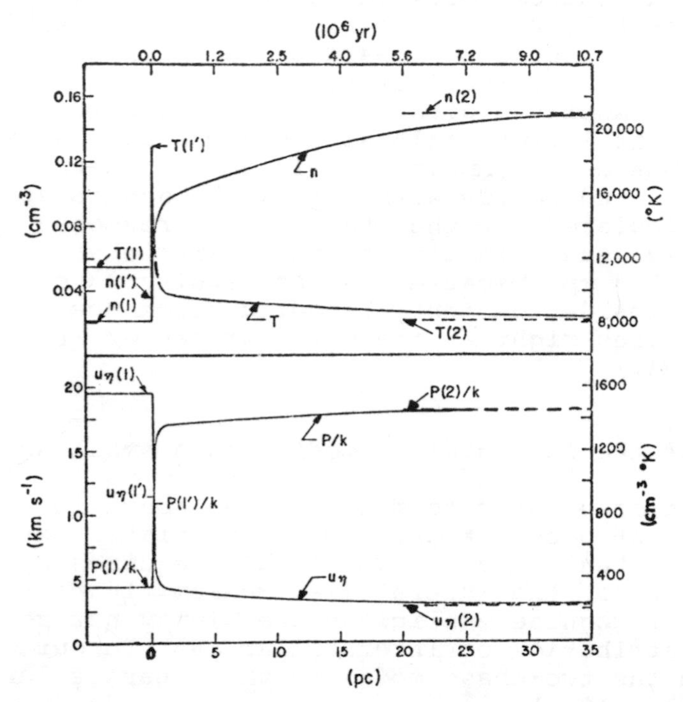

Fig. 2: The internal structure of the shock layer in the intercloud gas when all interactions with the cloud phase are ignored. (Based on corrected calculations of Shu et al., 1972)

phase. The abscissae show the distance normal to the shock front in parsecs and the time elapsed in the flow in millions of years. The horizontal solid lines show the incoming upstream values; the horizontal dashed lines, the asymptotic downstream values to which the gas must inevitably relax. The ticks show the values attained immediately after the viscous shock experienced by the gas.

The gas is raised to a high kinetic temperature as a result of the dissipation in the viscous layer, but it cools quickly by radiating away the shock-deposited thermal energy. The time spent to relax to the equilibrium appropriate to the downstream conditions is about 10^7 years. This is much shorter than the time spent between galactic shocks, about 2.7 x 10^8 years. Therefore, the treatment of the entire relaxation layer as a discontinuity for the large-scale flow is well-justified for the intercloud gas.

The situation for the cloud phase is more complicated. Massive interstellar clouds are decelerated much more slowly than individual gas atoms inside the shock layer because the deceleration occurs through the drag exerted between the two phases and the diffusive transport of momentum arising from cloud-cloud collisions. Moreover, the clouds will be collapsed simultaneously by the higher ambient pressure to denser internal states. Eventually, however, the cloud phase must also reach an equilibrium state consistent with the downstream conditions. The distance it takes to do this will be in the order of 10^2 pc.

2. Compressional history of the clouds in the galaxy. Shu et al. (1972) have discussed the implications of large-scale variations of the ambient gas pressure for the internal states of the interstellar clouds in the galaxy. Here, we shall follow their discussion which assumes solar values for the relative elemental abundances of the gas in the interiors of interstellar clouds. Should significant heavy-element depletion onto grains occur, even in relatively unobscured clouds, with very short characteristic timescales (Meszaros, 1973a, 1973b), then the details of the discussion given below may have to be revised considerably (see also Biermann et al., 1972).

If F = 5% at the solar circle as we believe, the total variation of pressure along a streamline exceeds the range of pressures consistent with the thermal stability of both cloud and intercloud phases, and phase transformations become inevitable. Although the detailed processes are probably quite complicated, the net effect of phase transitions can be discussed quite simply for the large-scale picture.

For a substantial portion of the "interarm" region, the internal densities of the clouds stay at relatively low values because the pressure cannot drop below a minimum value without some of the cloud material evaporating to maintain a constant pressure environment for the remaining clouds. Thus, the clouds between spiral arms would be so warm (temperatues in excess of 200°K) and so rarefied that they would quite difficultly be detected in 21-cm line absorption - in apparent agreement with the observations.

In a steady state the conversion of cloud material into intercloud material during the expansive phase of the flow is completely recovered subsequently in the compressive shock. Inside the shock layer, transient pressures higher than the maximum pressure, P_{max}, consistent with thermal stability of the intercloud medium force a conversion of some of the intercloud gas into the cloud phase. We have reason to believe that this conversion takes place near pre-existing clouds with the latter acting as "nucleation centers". In the large-scale picture, all of this is regarded as detail in a region taken to be infinitesimally thin, and the figures of Shu et al. (1972) merely show the aftermath of the phase transition when the gas pressure has dropped below P_{max} at point "2".

Shu et al. have also estimated that the compression by the galactic shock at the solar circle lowers the mass threshold for the gravitational collapse of interstellar clouds by a factor of about 25. If we have a reasonable mass distribution of interstellar clouds, such a lowering of the mass threshold would bring a substantial fraction of the clouds into collapse. (See Shu (1972) for a discussion of the processing rate of interstellar gas in such a picture.) This would go a long way toward explaining why the regions of substantial star formation can be delineated so sharply in certain external spiral galaxies. Of course, inclusion of the effects of transient pressures inside the galactic shock layer, cloud rotation, internal magnetic fields, internal turbulence, etc., may modify our numerical estimates substantially. However, we feel that our basic explanation of the sequence of events would be preserved.

REFERENCES

Biermann, P., Kippenhahn, R., Tscharnuter, W., Yorke,H., Astr. and Ap. 19, 113, 1972.

Contopoulos, G., Ap.J. 160, 113, 1970.

Dewar, R.L.A., Ap.J. 174, 301, 1972.

Feldman, S.I., Lin, C.C., Stud.Appl.Math., in press, 1973.

Field, G.B., Goldsmith, D.W., Habing, H.J., Ap.J.Lett. 155, L149, 1969.

Fujimoto, M., Proc.IAU Symp.No. 29, 453, 1966.

Hohl, F., Ap.J. 168, 343, 1971.

Hunter, C., M.N.R.A.S. 126, 299, 1963.

Kalnajs, A.J., Ph.D. Thesis, Harvard University, 1965.

Kalnajs, A.J., Ap.J. 166, 275, 1971.

Kalnajs, A.J., Ap.J. 175, 63, 1972.

Lin, C.C., Proc.IAU Symp. No. 38, 377, 1970.

Lin, C.C., in Highlights of Astronomy, Vol.II, p. 81, 1971.

Lin, C.C., Shu, F.H., Ap.J. 140, 646, 1964.

Lin, C.C., Shu, F.H., Proc.Nat.Acad.Sci. 25, 229, 1966.

Lin, C.C., Yuan, C., Shu, F.H., Ap.J. 155, 721, 1969.

Lynden-Bell, D., Kalnajs, A.J., M.N.R.A.S. 157, 1, 1972.

Lynds, B.T., Proc.IAU Symp. No. 38, 26, 1970.

Mark, J., Proc.Nat.Acad.Sci. 68, 2095, 1971.

Mathewson, D.S., van der Kruit, P.C., Brouw, W.N., Astr. and Ap. 17, 468, 1972.

Meszaros, P., Ap.J. 180, 381, 1973a.

Meszaros, P., Ap.J. 180, 397, 1973b.

Miller, R.H., Prendergast, K.H., Qurk, W.J., Ap.J. 161, 903, 1970.

Ostriker, J.P., Bodenheimer, P., Ap.J. 180, 171, 1973.

Peebles, P.J., Ostriker, J.P., in preparation, 1973.

Pikelner, S.B., Astr.Zh. 44, 1915, 1967.

Roberts, W.W., Ap.J. 158, 123, 1969.

Shu, F.H., Ph.D. Thesis, Harvard University, 1968.

Shu, F.H., Ap.J. 160, 89, 1970a.

Shu, F.H., Ap.J. 160, 99, 1970b.

Shu, F.H., Proc.IAU Symp. No. 52, in press, 1972.

Shu, F.H., in preparation, 1973.

Shu, F.H., Milione, V., Roberts, W.W., Ap.J., in press,1973.

Shu, F.H., Milione, V., Gebel, W., Yuan, C., Goldsmith, D.W., Roberts, W.W., Ap.J. 173, 557, 1972.

Shu, F.H., Stachnik, R.V., Yost, J.C., Ap.J. 166, 465,1971.

Spitzer, L., Scott, E.H., Ap.J. 158, 161, 1969.

Toomre, A., Ap.J. 139, 1217, 1964.

Toomre, A., Ap.J. 158, 899, 1969.

Toomre, A., Toomre, J., Ap.J. 178, 623, 1972.

Van der Kruit, P.C., Oort, J.H., Mathewson, D.S., Astr. and Ap. 21, 169, 1972.

Vandervoort, P.O., Ap.J. 161, 87, 1970.

Vandervoort, P.O., Ap.J. 166, 37, 1971.

Vandervoort, P.O., Ap.J. 180, 739, 1973.

Whitham, G.B., J.Fluid Mech. 22, 273, 1965.

Woodward, P., in preparation, 1973.

STAR FORMATION AND STAR DESTRUCTION

F.D. Kahn

Department of Astronomy, University of
Manchester, Manchester, Great Britain

I. STELLAR EVOLUTION

Before discussing problems of star formation and star destruction it will be as well to describe some of the more important features of stellar evolution. We begin at a very early stage, when an object has recognisably become a star. This means that the object is opaque and violent mass motions have ceased. It is supported by gas and/or radiation pressure against collapse under self-gravitation. In order of magnitude, the central pressure must then be equal to the product of the typical mass per unit area with the typical acceleration of gravity,

or $$P = \frac{M}{R^2} \times \frac{GM}{R^2} = \frac{GM^2}{R^4} \qquad (1.1)$$

Here P = pressure
M = mass of star
R = radius of star
and G = universal gravitational constant.

The contributions to the pressure are

radiation pressure $$P_{rad} = \frac{1}{3} aT^4 \qquad (1.2)$$

and gas pressure $$P_{gas} = \rho kT/m \qquad (1.3)$$

where

ρ = gas density $\approx M/R^3$,
T = temperature,
k = Boltzmann's constant,
a = Stefan's constant and
m = mean molecular weight.

We deduce from these relations that, in order of magnitude

$$P = \frac{GM^2}{R^4} = P_{rad} + P_{gas} = \frac{k}{m}\frac{M}{R^3} T + \frac{1}{3} aT^4 \qquad (1.4)$$

or, with

$$\vartheta = \left(\frac{ma}{3kM}\right)^{1/3} TR, \qquad \text{that}$$

$$\vartheta^4 + \vartheta = G\left(\frac{a}{3}\right)^{1/3}\left(\frac{m}{k}\right)^{4/3} M^{2/3} \qquad (1.5)$$

The two terms on the left-hand side of this equation derive, respectively, from radiation pressure and gas pressure. It is clear that the former dominates when the right-hand side is large, that is when the mass M is large. The latter dominates when the mass is small. The dividing line occurs roughly for stars with a mass of 100 $M_\odot$. We shall later discuss massive stars, where radiation pressure contributes significantly, in connection with the problem of compact HII regions. The present discussion deals with less massive stars, nearer to a solar mass or thereabouts. Here gas pressure is far more important.

Now consider such a star while thermal gas pressure dominates, but effects due to the Pauli exclusion principle are unimportant. In other words we neglect the contributions from Fermi pressure. We then have from equation (1.4) that

$$\frac{kT}{m} = \frac{GM}{R} \qquad (1.6)$$

In the early stages of the evolution of any star the radius R will be large. The internal temperature T will consequently be much too low for nuclear reactions to be possible. The energy radiated by the star must come from its gravitational self-energy W_{grav}. We therefore find for the luminosity L that

$$L = -\frac{d}{dt} W_{grav} = \frac{d}{dt}\frac{GM^2}{R} \qquad (1.7)$$

Under conditions of radiative equilibrium the flow of energy through a star is determined by the opacity. At the surface of the star the boundary condition is that the optical depth τ in one scale height H is

$$\tau = \varkappa \rho_s H \tag{1.8}$$

where $\varkappa$ is the opacity at the surface temperature T_e and surface density ρ_s.

A glance at some opacity tables (e.g. Cox and Stewart 1970) shows that it is very hard to get any reasonable amount of opacity in a gas below 3000°K, unless there is dust present. Therefore we get the Hayashi condition; the surface temperature of a star cannot be less than 3000°K (Hayashi,1966). This forces the star to be more luminous than it would be if its interior were in radiative equilibrium, and so the star has to be convective throughout. To obtain a time-scale for evolution along the Hayashi phase note that

$$L = \pi R^2 ac\, T_e^4 = \frac{d}{dt}\frac{GM^2}{R} \tag{1.9}$$

from which it easily follows that

$$R = \left(\frac{GM^2}{3\pi ac T_e^4}\right)^{1/3} t^{-1/3} \sim 10^{16} t^{-1/3} \tag{1.10}$$

for a star of solar mass. Thus the Hayashi contraction is most rapid early on, and slows down with time. The luminosity of the star also decreases as its radius shrinks. Eventually it reaches the stage at which radiative transport can bring energy to the surface at the rate demanded by the surface temperature T_e = 3000°K. The Hayashi phase then ceases and the star goes over to a phase of contraction under radiative equilibrium.

To find out how much energy is carried outwards per unit time we ought to do a detailed calculation involving the determination of the physical conditions and the opacity at various stages and in various places. This is too complex: instead we shall continue with our order-of-magnitude estimates, and shall use the time-honoured formula for the Rosseland mean opacity (Allen,1955) in which

$$\varkappa \propto \rho T^{-3.5} \propto \frac{M}{R^3}\,\frac{R^{3.5}}{M^{3.5}}$$

so that $\varkappa = qR^{0.5} M^{-2.5}$ say, (1.11)

where q is a suitable constant. In radiative equilibrium the luminosity of a star is, by order of magnitude, equal to the product of the energy flux per unit area with the surface area, divided by the optical depth to the centre, or

$$L = \pi R^2 \times ac\, T^4 \times \frac{1}{\kappa\rho R}$$

$$\propto R^2 \times \left(\frac{M}{R}\right)^4 \times \frac{M^{2.5}}{R^{0.5}} \times \frac{R^3}{M} \times \frac{1}{R} = M^{5.5} R^{-0.5} \qquad (1.12)$$

As usual we let $M_\odot$ be the mass of the sun, and $L_\odot$, $R_\odot$ respectively be the solar luminosity and radius on the main sequence (i.e. now) then

$$\frac{L}{L_\odot} = \left(\frac{M}{M_\odot}\right)^{5.5} \left(\frac{R_\odot}{R}\right)^{0.5} \qquad (1.13)$$

In terms of the surface temperature T_e of a star

$$\frac{L}{L_\odot} = \frac{R^2\ T_e{}^4}{R_\odot^2\ T_{e\odot}^4} \qquad (1.14)$$

A star has a surface temperature of 3000°K when it leaves the Hayashi phase and begins the radiative phase of contraction. Therefore, since $T_{e\odot} = 6000$°K we get

$$\frac{L}{L_\odot} = \frac{R^2}{R_\odot^2}\ \frac{1}{16} = \left(\frac{R_\odot}{R}\right)^{0.5} \qquad (1.15)$$

the second step here comes from equation (1.13), with $M = M_\odot$. It follows that the Hayashi phase ends, for a star of solar mass, when

$$R = (16)^{0.4} R_\odot \doteqdot 3\ R_\odot = 2 \times 10^{11}\,\mathrm{cm} \qquad (1.16)$$

According to equation (1.10) this stage is reached when

$$t = 10^{14}\ \mathrm{sec} = 3\ \text{million years.}$$

During the subsequent contraction towards the main sequence the luminosity increases by a factor of $\sqrt{3} \doteq 1.7$.

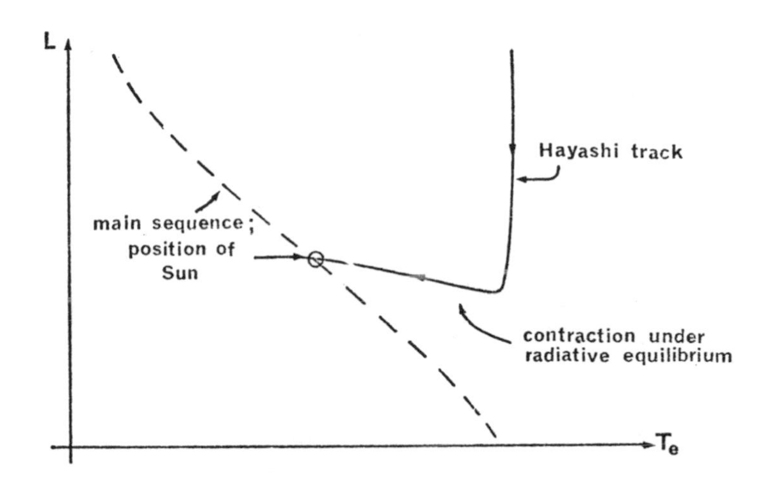

Fig. 1: Track of the evolution towards the main sequence in the L, T_e diagram.

Why does the sun's contraction - or that of any star- come to a halt on the main sequence? It can readily be calculated that the deep interior temperature of the sun on the main sequence is about 14 million $^{\circ}$K. At the end of its Hayashi phase, when its radius is about three times larger, the interior temperature of the sun would be only about 5 million $^{\circ}$K. Not many nuclear reactions will go at so low a temperature. In fact nuclear reactions only become important near the main sequence. In the sun and in most other stars the important reactions are those which convert H to He, either via the proton-proton chain (in low mass stars) at a rate, per unit mass, proportional to ρT^4, or via the CN-cycle (in higher mass stars) at a rate, per unit mass, proportional to ρT^{18}. The main sequence is the locus on which contraction is halted, for stars of different masses, because nuclear reactions generate enough energy by the conversion of H to He to balance the loss of energy through radiation from the surface.

With a law of energy generation, per unit mass,

$$Q \propto \rho T^n \propto \frac{M^{n+1}}{R^{n+3}} \tag{1.17}$$

we must have, on the main sequence,

$$\frac{L}{L_\odot} = \left(\frac{M}{M_\odot}\right)^{n+2} \left(\frac{R_\odot}{R}\right)^{n+3} \tag{1.18}$$

as well as the radiative condition (1.13)

$$\frac{L}{L_\odot} = \left(\frac{M}{M_\odot}\right)^{5.5} \left(\frac{R_\odot}{R}\right)^{0.5}$$

so that we get the mass-luminosity relation on the main sequence

$$\frac{L}{L_\odot} = \left(\frac{M}{M_\odot}\right)^{5\left(\frac{1+31/10n}{1+5/2n}\right)} \tag{1.19}$$

It makes little difference, with these formulae, whether $n = 4$ or $n = 18$; we get roughly that $L \propto M^5$ on the main sequence. But note that the position of the main sequence is somewhat dependent on the efficiency of the exothermic nuclear reactions that we invoke. If the nuclear reactions had been less efficient, then a higher internal temperature would have been needed to make them go fast enough to balance the radiative losses. Therefore contraction under radiative equilibrium would have proceeded to a small radius, and a higher luminosity. The main sequence would have been shifted thus:

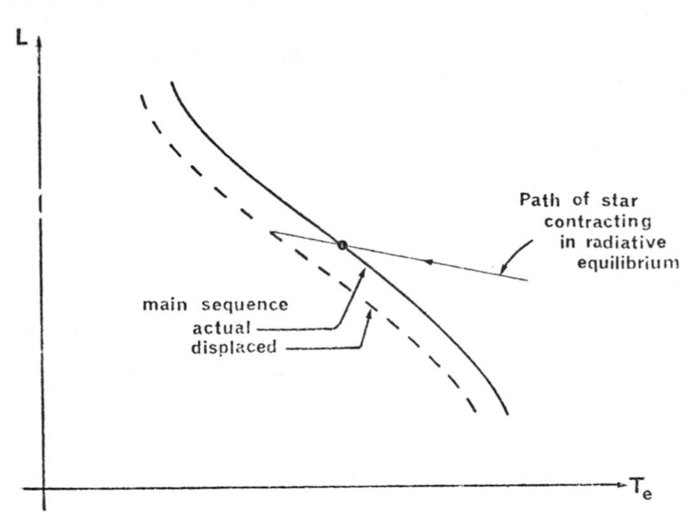

If there had been no nuclear reactions, the contraction would have just gone on, with continued release of gravitational energy. If endothermic nuclear reactions had occurred the contraction would have accelerated, and might have run away. We shall return to this point.

While a star is on the main sequence there will be changes in its chemical composition. When H is converted to He, energy is liberated at the rate of 6×10^{18}erg/g In the case of the sun a mass of 2×10^{33} gm liberates energy at a rate of 4×10^{33} erg/s; the rate of production of energy per unit mass is thus 2 erg/gm.s. In 10^{10} years, or 3×10^{17} seconds, the sun would convert about 10 per cent of its hydrogen content to helium, mainly in its deep interior where the temperature is highest. The composition of the sun becomes non-uniform, with a subsequent effect on its structure. It therefore begins to evolve away from the main sequence as in the following diagram:

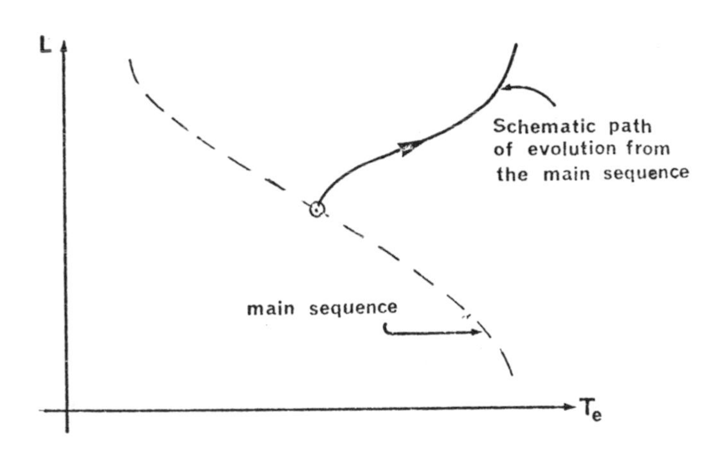

The evolution from the main sequence begins after about 10^{10} years, in the case of stars with the same mass as the sun. According to our earlier estimates L is proportional to M^5, and the rate of consumption of hydrogen goes like $\frac{L}{M}$, which is proportional to M^4. Thus evolution from the main sequence begins sooner for stars of greater mass. If we have a bunch of stars which we believe were all formed at about the same time, then we can determine the age of the bunch by the position of the "knee" in the Hertzsprung-Russell diagram, that is by finding the spectral type, or the surface temperature T_e, for those stars which are just beginning to evolve away from the main sequence. From the spectral type we infer the mass, and in this way we can deduce the age of the cluster to which the stars belong.

We now note that the stars in the galaxy do not form a homogeneous population. One commonly distinguishes between two populations thus:

Population I	Population II
Belongs to disk of galaxy.	Forms a spheroidal distribution.
Stars have approximately circular orbits in the disk.	Orbits of stars can be very eccentric.
Contains all the brightest (and therefore newest) stars.	Contains only old stars
Z ("heavy" element abundance by mass) ~ 0.02	Z ~ 0.0002.

The differences in chemical composition introduce differences in detail in the spectra of stars belonging to the two populations, and also in their modes of evolution. But the same principles apply, and one can make age determinations via cluster HR diagrams for Population II as well as Population I. It is then found that Population I clusters can have a range of ages from zero to almost the age of the galaxy (10^{10} years). Population II clusters are all about 10^{10} years old. The conclusion is that Population I has formed from the interstellar gas with its present spatial distribution and chemical composition. This composition seems to have changed little since -10^{10} A.D. + ε. But Population II must have formed from a gas whose composition was much nearer that expected of the intergalactic medium. At the time of its formation the interstellar gas was not concentrated towards a disk, but somehow we need to explain its state of motion. During the time that Population II stars were forming, or soon after, the interstellar gas was significantly enriched with heavy elements.

There is little good evidence that much enrichment has occurred since then. But theory suggests that nuclear transmutations which lead to a change in Z will occur as a star approaches collapse towards the supernova (SN) stage, and that the SN explosion then scatters the transmuted material through interstellar space. There is excellent evidence that many SN explosions have occurred since -10^{10} AD but not enough, it seems, to have significantly changed the value of Z.

Something should now be said about the late stages in the evolution of a star. The whole process of stellar evolution is an attempt by the star to come to thermodynamic equilibrium with space. In a sense the attempt is doomed to failure from the very start. As soon as the star forms, its interior becomes warmer than space, and heat leaks out. The star loses energy, contracts and becomes hotter inside. The process continues and eventually leads to interior temperatures that are high enough for nuclear reactions to occur, such as H → He, He → C, → O, ... → Fe. These exothermic reactions interrupt the contraction under gravity, but do not do so indefinitely. If there were no way of giving support against gravity other than by thermal pressure then sooner or later the supply of exothermically reacting substances would fail. After that there would be no way of forestalling the eventual catastrophic collapse. But under certain conditions a star can be supported by the pressure of a degenerate Fermi gas. The gas can consist either of electrons, in a white dwarf, or of neutrons in a neutron star.

Take the former case first, in a star for which the mean molecular weight (when fully ionized) is $\mu = \frac{4}{3} m_H$, m_H being the mass of the H atom. This applies to stars consisting of He^4, C^{12}, O^{16} ..... . There are $(2m_H)^{-1}$ electrons per gram in such a star, so that the electron density is $n_e \equiv \rho / 2m_H$, and the Fermi pressure is

$$P_F \sim \frac{h^2 n_e^{5/3}}{2m_e} \sim \frac{h^2 \rho^{5/3}}{2^{8/3} m_H^{5/3} m_e} \equiv \omega \frac{M^{5/3}}{R^5} \qquad (1.20)$$

Here ω is a suitable constant. The result is valid when the electrons are non-relativistic.

But we know that the pressure inside the star is given by

$$P = \frac{GM^2}{R^4} \; ;$$

the Fermi pressure P_F must be smaller than P, so that clearly

$$\omega \frac{M^{5/3}}{R^5} \leq \frac{GM^2}{R^4}$$

or

$$R \geq \frac{\omega}{GM^{1/3}} \qquad (1.21)$$

The limiting radius is reached, for a star of mass M, when there is equality in relation (1.21). At this stage the star is entirely supported by the Fermi pressure of the electrons, and thermal contributions to the pressure are negligible. This is a possible final state for a star; it is called a white dwarf.

But note that the typical energy of an electron is

$$\varepsilon \propto \frac{h^2 n_e^{2/3}}{2m_e} \propto \rho^{2/3} \propto \frac{M^{2/3}}{R^2} \tag{1.22}$$

In the terminal state of a white dwarf $R \propto M^{-1/3}$, so that $\varepsilon \propto M^{4/3}$. The electrons are clearly required to be more energetic inside white dwarfs with larger masses. For the more massive white dwarfs the Fermi gas becomes relativistic, and then

$$P_F \sim chn_e^{1/3} \times n_e = \omega' \frac{M^{4/3}}{R^4} \tag{1.23}$$

say. At this stage a final equilibrium state is possible only if

$$GM^2 = \omega' M^{4/3}$$

or

$$M = \left(\frac{\omega'}{G}\right)^{3/2} \tag{1.24}$$

This is Chandresekhar's limiting mass for a white dwarf, about 1.25 $M_\odot$. No star more massive than this can exist as a white dwarf.

But there is another possible state at higher density in which Fermi pressure plays an important role. This is found in neutron stars, where the electrons have a Fermi energy which is large enough for them successively to destroy the heavier nuclei, and eventually to push them into the protons to form neutron. The neutrons in turn then form a Fermi gas, and support the star against further contraction.

It is possible to estimate the range of masses for which neutron stars can exist. The details depend on the assumptions one makes about the equation of state at such high densities, but this is highly uncertain. The usual estimate is that no neutron star can have a mass larger than 0.5 or 1 $M_\odot$. The typical radius for a neutron star is 10 km. In its final state the gravitational binding energy is $\frac{GM}{10\text{km}}$ per unit mass, or about 10^{20} erg/gm. Some of this energy would be needed to undo the nuclear reactions that had previously occurred during the evolution of the star, and some energy might escape with a flood of neutrinos. In the case of a rotating neutron star some energy might also be emitted as pulsar radiation. But in general it seems that the formation of a neutron star can release up to 10^{53}erg on a very short time scale. Much of this energy would be available to power a SN explosion.

Note that the evolution towards a neutron star would begin at the centre of a star which is too massive to become a white dwarf. Its interior must also be hot enough to start endothermic reactions, notably, the destruction of heavy nuclei and the production of neutrino pairs. The neutrinos escape fairly easily, and the process speeds up. The conjecture is that the inner part of the star collapses, and if the collapsed section has a mass below the neutron star limit, then we expect a neutron star to be formed. If its mass is larger we would expect a black hole. In either case energy would be released in large amounts, and would blow off the outer part of the star.

But when a black hole forms at the centre of a star it is of course possible that the energy available for the explosion exceeds 10^{53}erg. This is because a black hole should have a mass larger than $0.5M_\odot$, and because its energy release can be as much as 4×10^{20} erg/gm.

II. CONDENSATION OF THE INTERSTELLAR GAS INTO A MORE COMPACT STATE

The discussion of this range of problems is based on the Virial Theorem, which will be established first. Consider a cloud of monatomic gas, with internal thermal and (possibly) mass motions. The cloud may contain gravitational, magnetic and/or electromagnetic fields. It is confined by an external pressure P_{ext}, which is held constant. The boundary of the cloud can move.

We enumerate the various contributions to the energy of the cloud. To find the kinetic energy, consider the cloud as a collection of atoms, a typical one having mass m_i and being at position $\underset{\sim}{r}_i$ at time t. This description will handle both the thermal and the mass motions. The kinetic energy becomes

$$U = \frac{1}{2}\sum m_i \, \dot{\underset{\sim}{r}}_i^2 \qquad (2.1)$$

The potential energy due to the various fields becomes, in obvious notation,

$$W = W_{grav} + W_{mag} + W_{e.m.} \tag{2.2}$$

There is also the potential energy, due to the volume V which the cloud occupies in the external pressure field,

$$W_P = P_{ext} V. \tag{2.3}$$

We next define a scale λ by setting

$$I \equiv \Sigma m_i \underset{\sim}{r}_i{}^2 = \lambda^2 \Sigma m_i \underset{\sim}{r}_i{}^{(o)2} \tag{2.4}$$

In this formula $\underset{\sim}{r}_i{}^{(o)}$ is the position vector for the ith particle at the time t_o. We can choose this instant to suit ourselves. Now put

$$\underset{\sim}{r}_i = \lambda \underset{\sim}{R}_i \tag{2.5}$$

As the particles move about, the vectors $\underset{\sim}{R}_i$ will vary, in general, but we nevertheless get that

$$\Sigma m_i \underset{\sim}{R}_i{}^2 = \frac{I}{\lambda^2} = \Sigma m_i \underset{\sim}{r}_i{}^{(o)2} = \text{constant} \tag{2.6}$$

Hence

$$\Sigma m_i \underset{\sim}{R}_i \cdot \dot{R}_i \equiv 0. \tag{2.7}$$

The kinetic energy of the cloud becomes

$$U = \frac{1}{2} \Sigma m_i \{(R_i \dot{\lambda})\}^2 = \frac{1}{2} \dot{\lambda}^2 \Sigma m_i \underset{\sim}{R}_i^2 + \frac{1}{2} \lambda^2 \Sigma m_i \dot{\underset{\sim}{R}}_i{}^2 \tag{2.8}$$

We note next that

$$W_{grav} = -\frac{1}{2} \sum_{i \neq j} \frac{G m_i m_j}{|\underset{\sim}{r}_i - \underset{\sim}{r}_j|} \equiv -\frac{1}{2\lambda} \sum_{i \neq j} \frac{G m_i m_j}{|\underset{\sim}{R}_i - \underset{\sim}{R}_j|}$$

W_{grav} therefore varies like λ^{-1}, as the scale of the cloud changes. The same form of variation applies to W_{mag}, if flux freezing holds, and to W_{em}, if changes are adiabatic. Finally W_p clearly scales like λ^3.

The Lagrangian of the system is

$$L = U - W = \frac{1}{2} \dot{\lambda}^2 \Sigma m_i \underset{\sim}{R}_i{}^2 + \frac{1}{2} \lambda^2 \Sigma m_i \dot{\underset{\sim}{R}}_i{}^2 - \{ W_{grav} + W_{mag} + W_{em} \} - W_p. \tag{2.10}$$

and the Lagrange equation for the generalized coordinate λ is

$$\frac{d}{dt}\left(\frac{\partial L}{\partial \dot{\lambda}}\right) - \frac{\partial L}{\partial \lambda} = 0 \qquad (2.11)$$

We now use our knowledge of the way in which the various quantities in L depend on λ or $\dot{\lambda}$, and find that (2.11) becomes, after some algebra,

$$(\lambda\ddot{\lambda} + \dot{\lambda}^2)\Sigma m_i \underset{\sim}{R}_i^2 = \lambda^2 \Sigma m_i \dot{\underset{\sim}{R}}_i^2 + \dot{\lambda}^2 \Sigma m_i \underset{\sim}{R}_i^2 + \frac{1}{\lambda}(W_{grav} + W_{mag} + W_{em})_{t=t_o} - 3\lambda^3 (W_p)_{t=t_o}$$

or

$$\frac{1}{2}\,\frac{d^2 I}{dt^2} = 2U + \Sigma W - 3W_p \qquad (2.12)$$

at time $t = t_o$; this is after all an arbitrary instant, and therefore equation (2.12) is valid in general. It states the Viral Theorem.

Now we specialise to consider a cloud which is in state equilibrium, has sperical symmetry, radius R, and contains no electromagnetic or magnetic fields. The equilibrium is simply a balance between pressure forces and gravitational fields, and demands that

$$2U + W_{grav} = 4\pi P_{ext} R^3 \qquad (2.13)$$

For an isothermal cloud at temperature T_c

$$U = \frac{3}{2}\frac{M}{m} kT_c, \qquad (2.14)$$

where m is the mean molecular weight. In order of magnitude

$$W = -\frac{GM^2}{R} \qquad (2.15)$$

and relation (2.13) becomes

$$\frac{3M}{m} kT_c = \frac{GM^2}{R} + 4\pi P_{ext} R^3 \qquad (2.16)$$

Suppose now that M, P_{ext} and T_c are given. Then equation (2.16) determines R. The right-hand side, regarded as a function of R, becomes large when R is small or R is large. It has a minimum when

$$R = R_{min} \equiv \left(\frac{GM^2}{12\pi P_{ext}}\right)^{1/4}, \qquad (2.17)$$

or

$$\frac{GM^2}{R^2} = 12\pi P_{ext} R^2 \qquad (2.18)$$

Therefore equilibrium can be attained only if

$$\frac{3MkT_c}{m} \geq \frac{GM^2}{R_{min}} + 4\pi P_{ext} R_{min}^3 \doteqdot 3(GM^2)^{3/4} P_{ext}^{1/4} \qquad (2.19)$$

or

$$\frac{kT_c}{m} \gtrsim G^{3/4} M^{1/2} P_{ext}^{1/4} \qquad (2.20)$$

The critical stage is reached when there is equality in (2.20), or when

$$\frac{M}{R^2} = \left(\frac{12\pi P_{ext}}{G}\right)^{1/2} \qquad (2.21)$$

and

$$\frac{kT_c}{M} = \frac{1}{(12\pi)^{1/4}} \frac{GM}{R} \approx 0.4 \frac{GM}{R} \qquad (2.22)$$

Just when the cloud will collapse depends on the mass per unit area, M/R^2, and on the temperature. But observation shows that these two quantities are physically related (Carruthers,1973). The critical quantity is the optical depth in the visual range due to the interstellar dust in the cloud. This is unity when M/R^2 is about 0.01 gm/cm^2. For surface densities below this value the inside of the cloud is illuminated by starlight, abundances of molecules are found to be low and the temperature will be relatively high, say 100°K. For higher surface densities the interior of the cloud will be dark, molecular abundances are high and the temperature will be low, say 10°K. The consequences can be shown schematically, in a diagram of log M versus log R (Fig.2).

We have seen that the external pressure determines the surface density M/R^2 at the point of collapse, and therefore the internal temperature. If the external pressure is low, the internal temperature is 100°K. The central isobar in the diagram separates the warm clouds (to the right and below) from the cold clouds (to the left and above). The critical M,R line therefore follows the 100°K isotherm, until it hits the central isobar. Its relevant portion is indicated by the double line. It terminates where

$$\frac{M}{R^2} = 10^{-2}\mathrm{gm/cm^2}, \quad \frac{kT_c}{m} = 0.4\,\frac{GM}{R}, \text{ with}$$

T_c = 100°K and m = 2 x 10^{-24}gm. This gives the values

$$M = 10^{37}\mathrm{gm}, \qquad R = 3 \times 10^{19}\mathrm{cm} \tag{2.23}$$

for the least massive cloud which can be brought to a critical condition on the warm sequence. Similarly the most massive cloud on the cold sequence has

$$M = 10^{35}\mathrm{gm} \text{ and } R = 3 \times 10^{18}\mathrm{cm} \tag{2.24}$$

at the point of collapse.

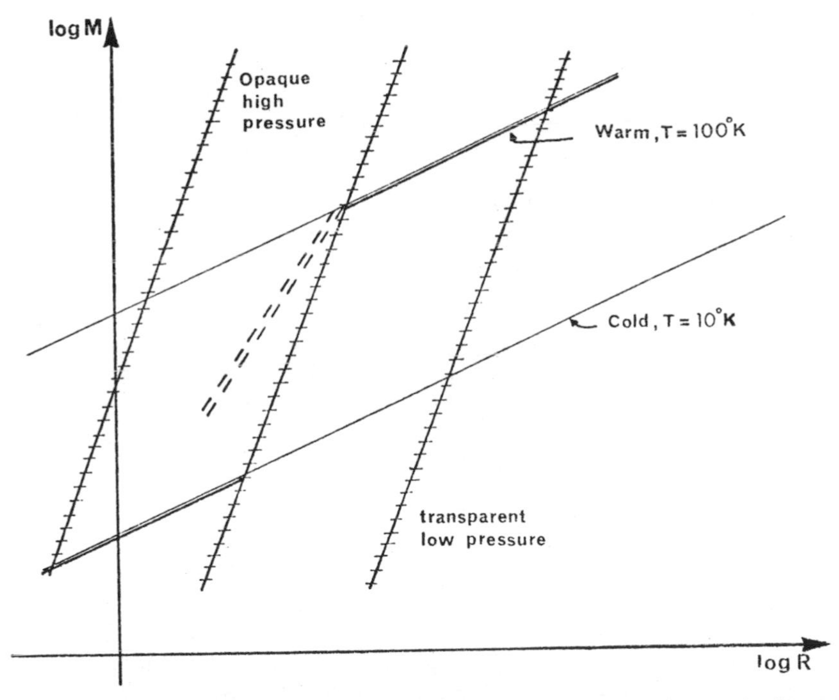

Fig. 2: Illustration of relations (2.21) and (2.22). R is the radius at which a cloud of mass M is brought to the point of collapse, under given conditions. Lines of equal pressure (isobars) are shown thus ┼┼┼┼┼┼┼┼ ; lines of equal temperature are shown ________.

Now consider how an interstellar cloud can be brought to the point of collapse. If its mass M is above 10^{37}gm, and the ambient pressure is low, then its radius will be large and the cloud will be stable. But if the external pressure is suddenly raised, perhaps by the passage of the cloud through a shock wave associated with a spiral arm, then the cloud may become compressed sufficiently to be made unstable. The necessary requirement is that it should be moved to a point in the log M-log R diagram above the warm sequence of critical states. If M is smaller than 10^{37} gm then the compression need not take the cloud all the way to the 100°K isotherm, since the cloud will become opaque (and therefore cool) sometime before it is compressed that far. The approximate location of the critical line has been indicated in the diagram with a double broken line.

Finally the cold sequence of critical states is, presumably, suitable for describing the collapse of more condensed clouds of interstellar gas, such as the globules that are often seen against H II regions.

It is worth noting that the typical collapse time $(R^3/GM)^{\frac{1}{2}}$ has a minimum value of 10^7 years on the warm sequence of critical states, and a maximum value of 3×10^6 years on the cold sequence. A time of 10^7 years is adequate for the evolution of a star whose mass exceeds 10 $M_\odot$. Since the density in a cloud decreases with radial distance from its centre, the actual collapse time will be shorter at the centre and longer at the outside. It would therefore be possible for a star, with mass greater than 10 $M_\odot$, to form and evolve at the centre of such a cloud before the collapse is complete. The H II region formed by this star, and perhaps the effects of a later SN explosion, should noticeably influence the evolution of the cloud in the later stages.

But only stars with masses greater than 30 $M_\odot$ can evolve in less than 3×10^6 years. It is therefore unlikely that a star should separate out and evolve during the collapse of a globule from the cold sequence of critical states. In the next chapter we shall briefly discuss the properties of cocoon stars. An object of this kind is probably formed after the collapse of a globule.

III. COCOON STARS

Recent observations have shown that a close connection exists between compact H II regions, infrared point sources and interstellar OH and H_2O masers (Mezger,1973, Habing,1973). It seems that all these phenomena are produced by a cocoon star. The light from the star, or the H II region is usually not observed, because of heavy obscuration by dust. This chapter gives a brief account of a physical model for a cocoon star. It is hoped soon to publish a more detailed paper on this subject.

First consider a mixture of gas and dust, falling freely under gravity into a star of mass M gm, at the rate $\dot{M}$gm/s. The speed of fall at distance R is $(2GM/R)^{\frac{1}{2}}$; in order to conserve mass in a steady flow the density must be

$$p(R) = \dot{M}(32\pi^2 GM)^{-1/2} R^{-3/2} \quad (3.1)$$

at distance R. The steady flow approximation is appropriate because the free fall time from a reasonable distance, say 10^{17} cm, is very short in comparison with any evolutionary time scale.

The stars involved in cocoons must be quite massive, say 50 $M_\odot$, in order to meet the energy budget required by observations. The details for the photon and energy output of such stars are easily calculated: we give some values in the table.

Mass: M		Luminosity: L	Lyman cont^m output: J	Evolution time scale
in $M_\odot$	in gm	(erg/s)	(photon/s)	(sec) t_{evol}
10	2×10^{34}	10^{38}	1.5×10^{48}	1.3×10^{14}
30	6×10^{34}	6×10^{38}	9×10^{48}	6×10^{13}
75	2.5×10^{35}	2.3×10^{39}	3.5×10^{49}	4×10^{13}

Reasonable approximations to these values are given by

$$L = 1.8 \times 10^{-17} M^{8/5}, \quad J = 2.7 \times 10^{-7} M^{8/5},$$

$$t_{evol} = 5 \times 10^{34} M^{-3/5} \quad (3.2)$$

with M in gms.

As the mixture of gas and dust comes nearer to the star the dust grains become warmer. There must be a distance R_m from the star at which an unshielded grain will melt. In this connection we need only consider the most refractory grains - i.e. graphite or silicate-;

the softer grains, e.g. those consisting of ice, would have been destroyed much further from the star. Our hypothetical unshielded grain is assumed to be conducting, and to have a radius of 0.035 μ. It absorbs radiation from the star, but re-radiates at a longer wavelength, characteristic of its temperature T_g. With our model we find that

$$T_g^6 = \frac{T_s^2}{ac} \frac{L}{4\pi R^2}; \quad (3.3)$$

T_s is a critical temperature, about 10^4 °K in the case of our grains. For the purpose of illustrating our argument we have assumed that the grains are destroyed when they are heated to 1500°K. This occurs at a distance

$$R_m = \left(\frac{L\ T_s^2}{4\pi\ acT_g^6}\right)^{1/2} = 6 \times 10^{-5}\ L^{1/2} \div 2.5 \times 10^{-13} M^{4/5} \quad (3.4)$$

from the star. The radiation pressure of light from the star there is $P_{rad} = 7 \times 10^{-4}$ dynes/cm. No grain can get closer to the star than distance R_m, since there will not be any other grains between it and the star to shield it from the stellar radiation. With our assumed value for T_g, R_m equals about 2×10^{15}cm for a 50 $M_\odot$ star.

What kind of H II region could possibly exist near the star, at distances $R < R_m$? There will be no extinction here, since the dust has gone. The net rate of recombinations, per cm^3, which destroy Ly-c photons is ~

$$2 \times 10^{-13} n_H^2 \sim 5 \times 10^{34}\ \ 2 \sim 2.5 \times 10^{39} \frac{\dot{M}^2}{MR^3} \quad (3.5)$$

with the help of (3.1) and some numerical substitutions. The number of H atoms + protons per cm^3 is denoted by n_H. If r_* is the radius of the star, and R_I the radius of the H II region, we get on balancing photon input and output, that

$$J = 2.7 \times 10^{-7} M^{8/5} = \int_{r_*}^{R_I} 2 \times 10^{-13} n_H^2 \times 4\pi R^2 dR$$
$$= 3 \times 10^{40} \frac{\dot{M}^2}{M} \log \frac{R_I}{r_*} \quad (3.6)$$

We can write this result in the form

$$R_I = r_* \exp\ (10^{-47}\ M^{13/5}/\ \dot{M}^2). \quad (3.7)$$

A reasonable value for r_* is 10^{12} cm. The H II region therefore remains well inside the dust free zone if the exponent in (3.7) is less than, say 5, or

$$\dot{M} > (2 \times 10^{-48} M^{13/5})^{1/2} \sim 4.5 \times 10^{21} \text{ gm/s} \qquad (3.8)$$

An H II region with so small a radius would suffer from heavy self-absorption at radio wavelengths, and might not be detectable. Certainly the compact H II regions that are observed seem to have dimensions much larger than even 10^{15}cm.

But it is very likely that there will be no H II region worth talking about while a heavy inflow mass continues. This is because the accretion flow should produce a false photosphere around the star. For radial distances less than R_m the gas flow will certainly be supersonic. It is therefore likely that shocks will exist in the flow - rather like white water exists in a mountain stream. Behind the shock the flow is subsonic, but it soon becomes supersonic again as the gas falls away towards the star. We envisage that the star is surrounded by a set of such shocked regions. For small enough values of R the compressed layer behind a shock may be thick enough to be opaque to the radiation from the star. It therefore forms a false photosphere. Preliminary estimates (for a star of 50 $M_\odot$) indicate that the false photosphere would be at radial distances of about 9 x 10^{13}cm and 2 x 10^{13} cm, when $\dot{M}$ equals 10^{24} and 10^{22} gm/s respectively. The effective temperatures corresponding to these values are 4000°K and 8000°K, and are much too low to permit the formation of an H II region. Only if $\dot{M}$ is as low as 10^{20} gm/s, or less, will the photosphere coincide with that of the star proper. Therefore one should not expect any H II region around a star which is accreting at a rate higher than 10^{20}gm/s.

We come finally to the lower boundary of the dusty region, just beyond radial distance R_m. Here the dust absorbs the radiation from the star, and takes up the radiation pressure. Friction against the gas transfers the mechanical effect to the flow as a whole. As long as the ram pressure π of the infalling gas exceeds P_{rad}, the gas can flow onwards towards the star even if the dust is destroyed. A simple calculation shows that

$$\pi = \frac{\dot{M}}{4 \quad R_m^2} \sqrt{\frac{2GM}{R_m}} \doteq 5 \times 10^{-26} \dot{M}, \qquad (3.9)$$

and this exceeds P_{rad} only if $\dot{M}$ exceeds 10^{22} gm/s.

With so high an accretion rate we should have a false photosphere around the star with an effective temperature of 8000°K, and no H II region to speak of. But when the accretion rate decreases below 10^{22} gm/s the radiation pressure on the dust becomes strong enough to drive a shock wave into the accreting gas flow (see diagram).

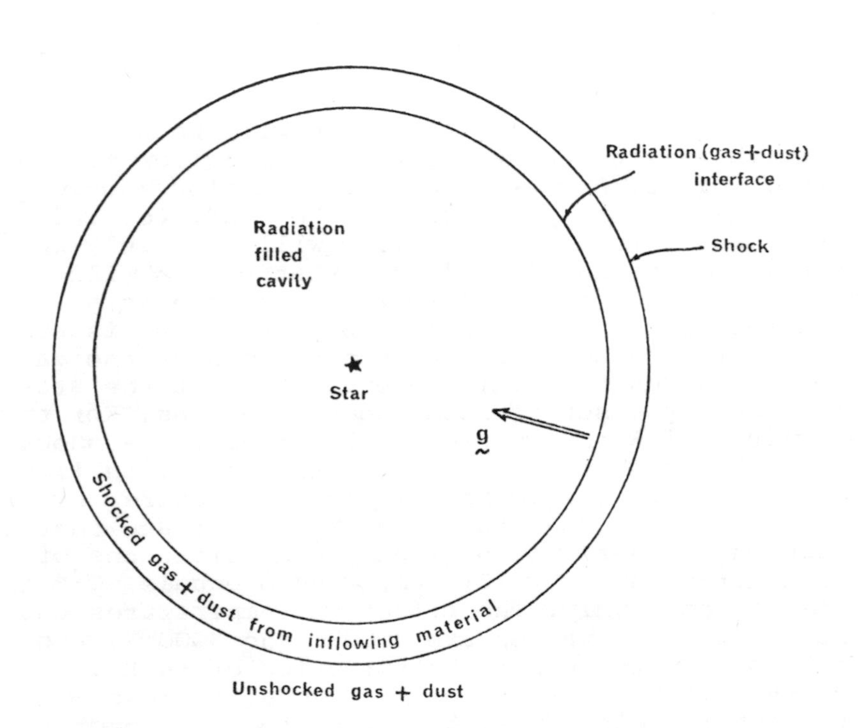

The resulting configuration is subject to the Rayleigh-Taylor instability, since a heavy fluid now rests, at the interface, on the light photon gas beneath. The shell should therefore break up into blobs, and tongues of the radiation filled region should be forced into the region further out. It must be at this stage of the evolution of a cocoon star that one observes the compact H II regions with their attendant phenomena.

IV SUPERNOVA REMNANTS

The best known supernova remnant (SNR) is the Crab Nebula. It originated in 1o54 AD and is now observed complete with a pulsar, and electromagnetic radiation output by synchrotron and by line emission. But the Crab

Nebula has been much discussed, and in any case it seems to be rather an extraordinary object. I shall therefore talk about Cassiopeia A instead.

This object is about 300 years old, shows a shell-like region where non-thermal radio emission occurs, and has a collection of filaments which emit in spectral lines at optical wavelengths. The polarization of the radio emission is generally perpendicular to the radial direction in the object. This implies that the radio emission takes place in a magnetic field, whose dominant direction is radial, and requires an explanation. So does the origin of the synchrotron electrons. It is not plausible to suggest that they originate in the SN explosion, since prohibitive losses of energy would occur in the expansion. A more reasonable explanation appears to be that the electrons are accelerated in the region where the gas expelled by the supernova meets the interstellar gas, and where turbulence most probably develops. The same turbulence also explains the radial direction of the magnetic field. (Much work has recently been done on this problem at Cambridge, notably by Gull, Longair, Rosenberg and Scheuer).

Let us now consider the early stages in the evolution of a supernova remnant.

The SN explodes. A mass M_* is given a large amount of energy, say 10^{18} erg/gm. A series of shocks criss-cross the star and lift off the mass M_*. The gravitational effects of the star make little difference to material with so high an energy content. The mass expands away from the star.

It would need a rather delicate calculation to discover how the kinetic energy distributes itself within the mass. We simplify our calculations by assuming that the density distribution within the mass is uniform and that the gas expands uniformly, once the thermal energy content has become small in comparison with the energy content due to the expansion velocity. Note that Gull has a different density distribution in his model of the explosion.

At time t the outer radius of the cloud of gas would be given by

$$R_o = V_i t \qquad (4.1)$$

if there were no deceleration because of the interstellar gas.

The kinetic energy of the cloud is

$$U = \frac{1}{2} M_* V_i^2 \frac{\int_0^1 x^4 dx}{\int_0^1 x^2 dx} = \frac{3}{10} M_* V_i^2 \tag{4.2}$$

Since the speed is xV_i at radial distance xR_o. If $\frac{U}{M_*} = 10^{18}$erg/gm, $V_i = 1.6 \times 10^9$ cm/s. This seems a reasonable value to take for the purpose of illustrating the argument.

The gas expelled by the SN runs into the interstellar medium, for which we assume a density $\rho_o = 2 \times 10^{-24}$ gm/cm^3. Therefore a shell develops; it is bounded on the outside by a shock advancing into the undisturbed interstellar medium, and on the inside by a shock moving inwards relative to the outflowing gas (see following diagram).

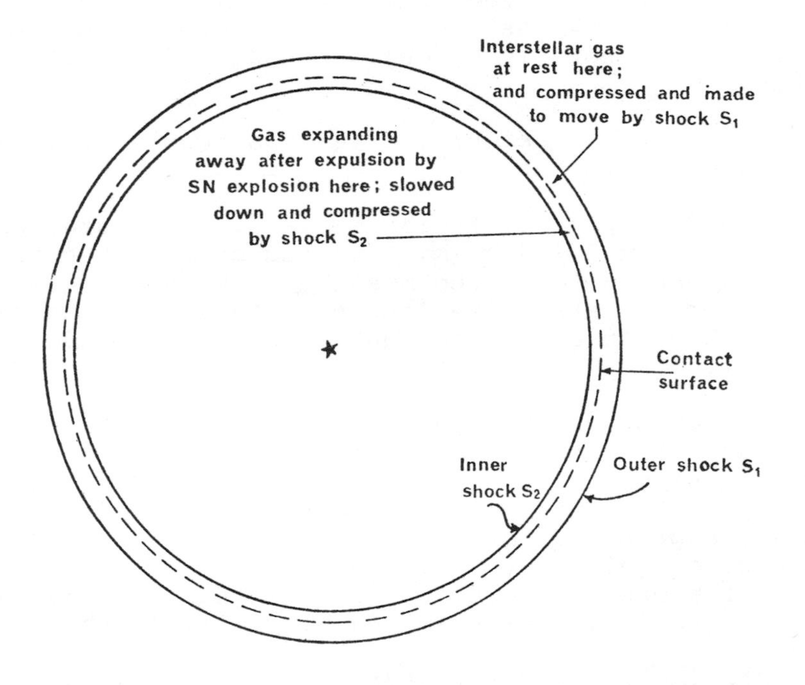

The gas in interstellar space is cold before shock S_1 overtakes it. So is the gas from the SN before it runs into shock S_2. Therefore both shocks are strong. The shocked gas is, in each case, compressed by a factor 4. At time t let the contact surface have radius $R (< V_i t)$.

The shocked gases form a thin shell around this radius.

The contact surface moves out with speed $\dot{R}$. In reasonable approximation shock S_1 also advances at speed $\dot{R}$ into the interstellar gas. But shock S_2 has a velocity $\dot{R} - \frac{R}{t}$) relative to the gas coming from the supernova.

Both shocks will be strong, since the interstellar gas and the gas from the star are both cold. The pressure in the shocked gas behind S_1 is therefore

$$P_1 \doteqdot \frac{3}{4}\rho_o\dot{R}^2 \doteqdot \frac{3}{4}\rho_o V_i^2 \tag{4.3}$$

and the pressure in the shocked gas behind S_2 is

$$P_2 \doteqdot \frac{3}{4} \times \frac{3M_*}{4\pi V_i^3 t^3}(\dot{R} - \frac{R}{t})^2 \tag{4.4}$$

We must now allow for the deceleration of the shell due to its interaction with the interstellar gas, and set

$$R = V_i t\left\{1 - \varepsilon(t)\right\} , \tag{4.5}$$

where $\varepsilon(t)$ is small compared with unity. It turns out that $\varepsilon(t)$ must have the form $\eta\, t^{3/2}$, and so we get

$$\dot{R} = V_i(1 - \frac{5}{2}\eta t^{3/2}) \tag{4.6}$$

and

$$\ddot{R} = -\frac{15}{4}\eta V_i t^{1/2} \tag{4.7}$$

The expression for the pressure behind S_2 becomes

$$P_2 = \frac{9}{16\pi}\frac{M_* V_i^2}{V_i^3 t^3} \times \frac{9}{4}\eta^2 t^3 = \frac{81}{64\pi}\frac{M_*}{V_i}\eta^2 \tag{4.8}$$

The total mass of gas in region 2 of the shell, at time t, is

$$M_2(t) = M_*\left\{1 - (\frac{R}{V_i t})^3\right\} \doteqdot 3 M_*\eta\, t^{3/2} \tag{4.9}$$

The total mass of gas in region 1 at the same time is

$$M_1(t) = \frac{4\pi}{3}\rho_o V_i^3 t^3 \tag{4.10}$$

We define a scale time t_s which is such that a mass M_* of interstellar gas would have been swept up (by the undecelerated shell) when $t = t_s$. Therefore

$$t_s^3 = \frac{3M_*}{4\pi\rho_o V_i^3} \tag{4.11}$$

Note that

$$M_1(t):M_2(t) = \frac{4\pi}{q} \quad \frac{\rho_o V_i^3}{\eta M_*} t^{3/2} : 1, \tag{4.12}$$

so that M_2 much exceeds M_1 when t is small, in the early phases of the evolution. We take the mass of the shell to be given by (4.9) at this stage. The deceleration of the shell is caused by the difference between the pressure P_1 and P_2, and we have that

$$\frac{M_2(t)}{4 \quad R^2} \ddot{R} \doteq - P_1 + P_2$$

or

$$\frac{3M_*\eta t^{3/2}}{4\pi V_i^2 t^2} \left(-\frac{15}{4}\eta V_i t^{1/2}\right) = -\frac{3}{4}\rho_o V_i^2 + \frac{81}{64\pi} \frac{M_*\eta^2}{V_i} \tag{4.13}$$

This simplifies to give

$$\eta^2 = \frac{16\pi}{87} \frac{\rho_o V_i^3}{M_*} \equiv \frac{4}{29} t_s^{-3} \tag{4.14}$$

Hence

$$R = V_i t \left\{ 1 - 0.38 \left(\frac{t}{t_s}\right)^{3/2} \right\} \tag{4.15}$$

$$\ddot{R} = - 1.4 \frac{V_i t^{1/2}}{t_s^{3/2}} \tag{4.16}$$

$$M_2 = 1.1 M_* \left(\frac{t}{t_s}\right)^{3/2} \tag{4.17}$$

and

$$P_2 = 0.23 \rho_o V_i^2 \tag{4.18}$$

If cooling of the shocked gases by radiation is negligible, then the densities P_1 and P_2 are easily found to be

$$\rho_1 = 4\rho_o \text{ and } \rho_2 = \frac{4M_*}{\frac{4}{3}\pi V_i^3 t^3} = \frac{3M_*}{\pi V_i^3 t^3}$$

so that

$$\rho_1 : \rho_2 = \frac{4\pi}{3} \frac{\rho_o V_i^3 t^3}{M_*} : 1 = t^3 : t_s^3 \quad (4.19)$$

The temperature T_1 is given by

$$P_1 = \frac{4k}{m} \rho_o T_1 = \frac{3}{4} \rho_o V_i^2$$

with $m = 10^{-24}$ gm for a fully ionized gas. In the present model $V_i = 1.6 \times 10^9$ cm/s and we get

$$T_1 \doteqdot 3 \times 10^9 \ {}^oK. \quad (4.20)$$

The gas in part 1 of the shell is therefore so hot that it can only cool at a completely negligible rate. But the gas in region 2 has a lower temperature T_2, given by

$$\frac{k}{m} \rho_2 T_2 = \frac{k}{m} \rho_1 T_2 \left(\frac{t_s}{t}\right)^3 = 0.23 \rho_o V_i^2$$

so that

$$T_2 \doteqdot 10^9 \left(\frac{t}{t_s}\right)^3 \ {}^oK, \quad (4.21)$$

at time t.

When the shell decelerates, the hydrostatic equilibrium within it sets itself as though there were a gravitational acceleration

$$g_{eff} \equiv |\ddot{R}| = 1.4 \frac{V_i t^{1/2}}{t_s^{3/2}}$$

directed radially outwards. The equilibrium is clearly Rayleigh-Taylor unstable, since a dense gas in region 2 lies over a light gas in region 1.

In the description of Cass A, given by Gull, energy can be extracted from this unstable configuration. We can estimate how much is available, as follows. A bubble of gas from region 1 expands, when it enters region 2, because the pressure there is lower by a factor of order 3. If the expansion is adiabatic, the density decreases by a factor $3^{3/5} \sim 2$. Hence the density ratio between the bubble and the ambient medium is $\rho_{1/2}\ \rho_2$, the buoyancy force per gram is $2\rho_2|\ddot{R}|/\rho_1$ and it acts over the thickness ΔR_2 of part 2 of the shell. The potential energy which can be extracted by moving the material of shell 1 through that of shell 2 is therefore

$$W = M_1 \Delta R_2 \times \frac{2\rho_2}{\rho_1} |\ddot{R}| = \frac{2M_1}{\rho_1} \frac{M_2}{4\pi R^2} |\ddot{R}| = 2M_2 \Delta R_1 |\ddot{R}| \quad (4.22)$$

and since

$$\Delta R_1 \sim \frac{R}{12} \sim \frac{V_i t}{12}$$

we get, with the help of (4.16) and(4.17) that

$$W = 2 \times 1.1\, M_* \left(\frac{t}{t_s}\right)^{3/2} \times \frac{V_i t}{12} \times 1.4\, \frac{V_i t^{1/2}}{t_s^{\,3/2}} \doteq 0.25\, M_* V_i^2 \left(\frac{t}{t_s}\right)^3 \quad (4.23)$$

The amount of energy that is available depends on the time t. We note for comparison that the total kinetic energy of the expanding supernova remnant at injection is $0.3\ M_* V_i^2$. If our formula is valid to about time $\frac{1}{3} t_s$, then it seems possible to release at least a few per cent of the injection energy, which was about 10^{52} erg in this model. Interpretations of the radio observations require an energy, in magnetic fields and electrons, of the order of 10^{49} erg. This should be attainable with a reasonable efficiency for the conversion of the potential energy in the shell. Note that the relative motions of gas in the shell will be mainly in a radial direction. Therefore the magnetic field lines should also tend to be stretched in a radial direction.

Finally, let us consider the formation of the fast filaments. The cooling time t_c for the gas in part 2 of the shell can be quite short. In his recent review article, Woltjer (1972) states that the rate of loss of energy by a gas with a temperature of the order of 10^6 °K is $2.5 \times 10^{-16} n_H/T$ erg per atom per second. In region 2 the energy content per atom is

$$3kT_2 = 4 \times 10^{-7} \left(\frac{t}{t_s}\right)^3 \ {}^{\circ}K \quad (4.24)$$

with the help of formula (4.21). The cooling time becomes

$$t_c = \frac{3kT_2^{\,2}}{2.5 \times 10^{-16} n_{H,2}} \quad (4.25)$$

Now

$$2 \times 10^{-24} n_{H,2} = P_2 = \frac{3M_*}{\pi V_i^3 t^3}$$

and, on using 4.25 and substituting numerical values we find that

$$t_c = 5 \times 10^{17} \left(\frac{t}{t_s}\right)^9 .$$

Cooling can be effective as long as $t_c < t$, or

$$t < 6 \times 10^{-3} t_s^{9/8} .$$

With our numerical values, $t_s = 6 \times 10^9$ s and cooling becomes ineffective after time

$$t = \frac{10^{11}}{160} \text{sec} \sim 6 \times 10^8 \text{ sec} \sim 20 \text{ yrs},$$

At this time the mass in part 2 of the shell is

$$1.1 M_* \left(\frac{t}{t_s}\right)^{3/2} \doteqdot 0.04 M_* \sim 4 \times 10^{32} \text{ gm} \qquad (4.26)$$

This is our estimate for the total mass that can possibly be formed into filaments. It seems adequate. It also seems clear that the filaments will become quite dense when they have cooled down to 10^4 °K, and that they will be able to maintain, for a long time, the high speed which they had at the epoch when they were formed.

We finally make an estimate for the lifetime of a synchrotron electron. Rosenberg (1970) quotes a typical magnetic field strength of $5 \times 10^{-4} \Gamma$. At 5 GHz the value of ω is 3×10^{10} s^{-1}, and therefore the value of γ for the synchrotron electrons is given by

$$3 \times 10^{10} \sim \gamma^2 \frac{eH}{mc} \sim 10^4 \gamma^2$$

so that $\gamma \sim 2000$.

Such an electron has a lifetime

$$\frac{\gamma mc^2}{\gamma^2 e^4 H^2/m^2c^3} = \frac{m^3c^5}{\gamma e^4 H^2} \doteqdot 10^{12} \text{ sec} = 30\,000 \text{ yrs}$$

with respect to synchrotron losses. It is therefore entirely possible that the electrons which we observe now (when Cass A is 300 years old) were produced when Cass A was only 100 years old. We make this remark because $t_s \sim 300$ years with the numerical values adopted for our model, and because it seems that the acceleration of the synchrotron electrons takes place most efficiently some time before $t = t_s$.

V. COLLAPSE OF THE PROTO-GALAXY

In this lecture we shall briefly consider some simple mechanical aspects of the formation of the galaxy itself. We want to see how a mass of gas of about $10^{11} M_\odot$ can be brought to the point of collapse under self-gravitation. At the time when the formation of the galaxy took place there were, presumably, not many stars around to heat the diffuse gas. We therefore assume that the thermal energy of the gas was derived from its gravitational self-energy. Further the intergalactic gas is usually thought to be very poor in the heavier elements. Therefore the rate of cooling by dust and molecules must also have been negligible. Cooling would be due mainly to radiation by ionized hydrogen, at a rate

$$L_H \doteqdot 10^{-3} n_H T^{1/2} + 300 n_H T^{-1/2} (\text{erg gm}^{-1}\text{s}^{-1}) \quad (5.1)$$

for a fully ionized gas with n_H protons/cm^3 (Minkowski, 1942). When the temperature exceeds 3×10^5 °K the second term on the right-hand side becomes negligible. This is so in the present case. The characteristic cooling time is therefore

$$t_c = \frac{\frac{3k}{m} T}{10^{-3} n_H T^{1/2}} \doteqdot 2 \times 10^{11} \frac{T^{1/2}}{n_H}, \quad (5.2)$$

if the mean molecular weight is $m = 2 \times 10^{-24}$gm.

Let us model the proto-galaxy by a polytrope with central pressure P_*, density ρ_*, and index $n = 3/2$. The scale length for this model is

$$l = \left\{ \frac{(n+1) P_*}{4\pi G \rho_*^2} \right\}^{1/2} = \left(\frac{5}{8\pi G} \right)^{1/2} \frac{P_*^{1/2}}{\rho_*} \quad (5.3)$$

The characteristic dynamical time at the centre is

$$t_D \sim \frac{l}{(5P_*/3\rho_*)^{1/2}} = \left(\frac{3}{8\pi G \rho_*} \right)^{1/2} \quad (5.4)$$

The centre will collapse if significant cooling can occur in a time shorter than the dynamical time, that is if $t_C < t_D$ or

$$\frac{2\times 10^{11} T_*^{1/2}}{n} \equiv \frac{2\times 10^{11} T_*^{1/2}}{5\times 10^{23}\rho_*} \equiv 4\times 10^{-13}\ \frac{T_*^{1/2}}{\rho_*} \lesssim \left(\frac{3}{8\pi G\rho_*}\right)^{1/2} \qquad (5.5)$$

The inequality can be re-written in the form

$$l^2 = \frac{5P_*}{8\pi G\rho_*} \lesssim 6\times 10^{24}\frac{k}{m}\left(\frac{15}{8\pi G}\right)^2 = 2.5\times 10^{45}\,\mathrm{cm}^2$$

or

$$l \lesssim 5\times 10^{22}\,\mathrm{cm} \sim 16\,\mathrm{kpc} \qquad (5.6)$$

This result is independent of the mass of the proto-galaxy. Half the mass of a polytrope of index $^3/2$ lies within a distance $R_{1/2} = 2\,\ell$ of the centre. Therefore the typical linear scale is $\sim$ 30 kpc when collapse begins.

The temperature T_* at the centre depends on the mass of the proto-galaxy through the relation

$$\frac{2kT_*}{m} = \frac{P_*}{\rho_*} \sim \frac{GM_*}{R_{1/2}}$$

This leads to an estimate of $T \sim 10^6\ {}^\circ K$. The typical time scale for the collapse to a high density is

$$t_{D*} \sim \left(\frac{3}{8\pi G\rho_*}\right)^{1/2} \qquad (5.7)$$

for the material at the centre. For material which begins the collapse from distance

$$r \equiv l\varkappa$$

the typical time scale is

$$t_D \sim \left(\frac{3}{8\pi G\bar{\rho}}\right)^{1/2} \qquad (5.8)$$

where

$$\bar{\rho} = \rho_*\,\frac{\int_0^{\varkappa} U^{3/2}\xi^2 d\xi}{\int_0^{\varkappa}\xi^2 d\xi} \qquad (5.9)$$

is the mean density within a sphere of dimensionless radius x.

In the inner part of any polytrope the dimensionless gravitational potential is approximated by

$$U \doteq 1 - \frac{x^2}{6}$$

so that

$$\bar{\rho} \doteq \rho_* \left(1 - \frac{3x^2}{20}\right)$$

and

$$\tau(x) \equiv t_D - t_{D*} \doteq \frac{3}{40}\left(\frac{3}{8\pi G\rho_*}\right)^{1/2} x^2 \tag{5.10}$$

The polytrope contains a mass $M(x) =$

$$\frac{4\pi}{3}\rho_* l^3 x^3 = \frac{4\pi}{3}\left(\frac{5}{8\pi G}\right)^{3/2} \frac{P_*^{3/2}}{\rho_*^2} x^3 \tag{5.11}$$

out to dimensionless radius x. A mass $M(\tau)$ collapses to the centre in the time interval τ after a singularity first forms. It is given by

$$M(\tau) = \frac{4\pi}{3}\left(\frac{40}{3}\right)^{3/2}\left(\frac{25}{24\pi G}\right)^{3/4} \frac{P_*^{3/2}}{\rho_*^{5/4}} \tau^{3/2} \tag{5.12}$$

With substitutions

$$\frac{P_*}{\rho_*} \sim \frac{GM_*}{R_{1/2}}, \quad \rho_* \sim \frac{M}{R_{1/2}^3}$$

we get from (5.12), that

$$M(\tau) \doteq 10^3 \left(\frac{G}{8\pi}\right)^{3/4} \frac{M^{1/4}}{R_{1/2}^{9/4}} \tau^{3/2} \tag{5.13}$$

But we expect a rather intense period of supernova explosions to begin at about time $\tau \sim 10^{14}$ seconds, when the massive stars at the galactic centre have completed their full evolution. These will, presumably, send shock waves out into the proto-galaxy, and halt the collapse. At time $\tau = 10^{14}$ seconds, M (τ) equals about 10^{43} gm or $5 \times 10^9 M_\odot$; this is a value obtained for a proto-galaxy whose mass is $10^{11} M_\odot$.

It therefore seems likely that the initial burst of star formation will have involved a mass of about $5x10^9 M_\odot$ at the galactic centre. This took place very quickly, and perhaps led to the production of the heavy elements in the galaxy. The formation of Population II stars must then have occurred in the protogalaxy after it was disturbed by the SN explosions at the galactic centre. Population I stars would have formed rather later, when all these disturbances had died away, and the heavy elements had become well mixed with the gas in the galaxy.

BIBLIOGRAPHY

I. The evolution of protostars is described by C. Hayashi in Ann. Rev. Astron. Astrophys. 4, 171, 1966. For a different view of the process, see the papers by R.B. Larson in Monthly Not. Royal Astron. Soc. 157, 121, 1972, and in the proceedings of the Symposium on the Origin of the Solar System (Nice 1972) p.142. The old formula for the Rosseland mean is quoted by C.W.Allen in "Astrophysical Quantities" (Athlone Press, London,1955) p. 93. More up to date results are given by A.N. Cox and J.N. Stewart in Ap.J. (Supplements) 19, 243, 1970.

Various aspects of stellar structure are discussed in volume 8 of Stars and Stell. Syst., notably

B. Strömgen on " Main Sequence Stars"

H. Reeves on "Stellar Energy Sources"

R.L. Sears and R.R. Brownlee on "Stellar Evolution and Age Determinations",

L. Mestel on "White Dwarfs"

F. Zwicks on "Supernovae"

S. Bashkin on "Origin of Elements".

We also refer to a review by O.J. Eggen on "Observational Aspects of Stellar Evolution" in Ann. Rev. Astron. Astrophys. 3, 235, 1965 and by A.Weigert on "Stellar Evolution According to Numerical Methods" in "La Structure Interne des Etoiles", Saas Fee meeting 1969. A.G.W. Cameron reviews "Neutron Stars" in Ann. Rev. Astron. Astrophys. 8, 179,1970.

II. Problems of star formation are discussed by the various contributors to "Die Entstehung von Sternen", Springer 1960, and by Lyman Spitzer in his chapter on

"Dynamics of Interstellar Matter, and the Formation of Stars",in volume 7 of Stars and Stell. Syst. See also R.B. Larson "Collapse of a Rotating Cloud"in Monthly Not. Royal Astron. Soc. 156, 437, 1972.

The relation between the optical depth and probable temperature of a cloud was discussed by R.G. Carruthers in his lectures at this meeting.

III. P.G. Mezger and H.J. Habing reviewed the observations on compact H II region at this meeting. The overall optical properties of grains in the infrared, visible and U.V. are described by J.M. Greenberg in Astron. and Astrophys. 12, 240, 1971.

IV. For a recent review on Supernova remnants see L. Woltjer Ann. Rev. Astron. Astrophys. 10, 129, 1972. Observations on Cass A at radio wavelengths are discussed by I. Rosenberg in Monthly Not. Royal Astron. Soc. 151, 109, 1970. S.F. Gull first described the mechanism for extracting energy from the stratifaction of the supernova shell in Monthly Not. Royal Astron. Soc. 161, 47, 1973.

V. O.J. Eggen, D. Lynden-Bell and A.R. Sandage discuss the observational evidence concerning the early stages of the galaxy in Ap. J. 136, 748, 1962. An interesting model of the collapse of the proto-galaxy is described by S. Grzedzielski in Monthly Not. Royal Astron. Soc. 134, 109, 1966. The formula for the rate of cooling of the gas was originally given by R. Minkowski Ap. J. 96, 199, 1942.

THE NUCLEOSYNTHESIS OF THE LIGHT ELEMENTS AND THE INTERSTELLAR MEDIUM

H. Reeves

Institut d'Astrophysique, Paris, France

I. INTRODUCTION: WHERE, WHEN AND HOW; THE BUILD-UP OF AN EAC

At any one time the chemical composition of the interstellar medium is a reflection of the time - integrated past history of the combined formation and destruction rate of the various chemical elements or isotopes.

Each element and/or isotopes will be followed through its Evolutionary Abundance Curve (EAC) in the interstellar medium. Samples of the interstellar matter can be analysed either through observations of the interstellar gas today (t=0), or through surface abundances of stars of various ages if we convince ourselves that these abundances have not been altered since the birth of the star. Of great interest to us will be the solar system (earth-moon-meteorites) since it provides a lot of data on the chemical composition of the gas some 4.6×10^9y ago ($t=t_\odot$).

The EAC of any one element will depend on the formation mechanisms and on destruction mechanisms. For convenience, elements belonging to same formation mechanisms (as far as we know) will be discussed together as a group. Therefore we start by discussing the major phases of nucleosynthesis. In Fig. 1 five possible phases are represented on a time axis. The first phase occupies some twenty minutes following the birth of the universe. (I assume here that there was such a thing as a big bang, as point of view that most-but-not all astrophysicists are ready to take).

Next is a hypothetical phase which would have preceded the birth of the galaxy (pre-galactic) and finally three phases taking place during the galactic life itself and defined here according to the mean energy of the nuclear processes involved in the element formation. The higher energy phase (~GeV) refers to Galactic Cosmic Rays, the (still hypothetical) medium energy phase (~10 MeV) refers to Shock Waves in Supernovae, and the low energy phase (keV-MeV) to Stellar Nucleosynthesis.

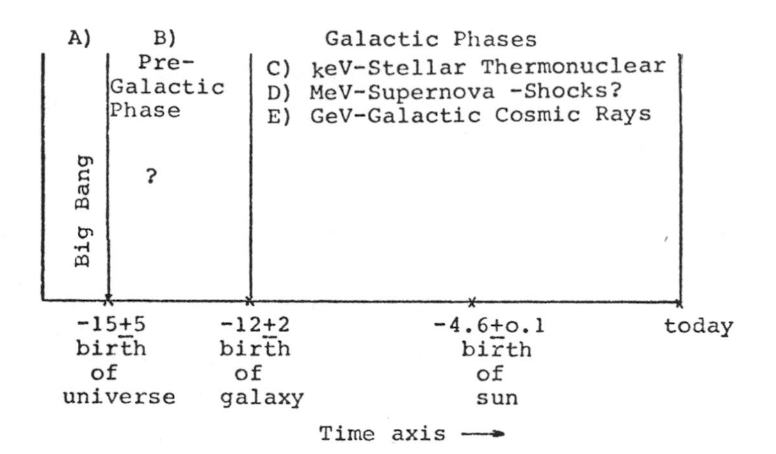

Fig. 1: Major phases of nucleosynthesis.

II. BIG BANG NUCLEOSYNTHESIS

The past history of the universe can be reconstituted (as far as the bulk thermodynamical properties are concerned) if we accept general relativity. For nucleosynthesis the most important relation is simply the constancy of the number of photons per nucleons

$$n_\gamma / n_N = \text{cst} \quad \text{or} \quad \rho_N \propto T_\gamma^3$$

This expression is almost always true. We know when it is not true (for instance when electron-positron annihilation takes place) and we can make the appropriate corrections. Hence, if we know ρ_N and T_γ today we know them at all times and, in particular, we know them when the universe was hot enough for nuclear reactions to take place. Hence we can compute the result of big bang nucleosynthesis (Peebles, 1971; Wagoner et al., 1967; Wagoner, 1973).

We think we know T_γ : the microwave observations of a universal black body radiation at 2.7 K is a good candidate for being the remnant of the past glory of the big bang (Fig. 2).

Assuming that general relativity is valid you can make computations with very few free parameters.

Big Bang provides simple explanations for the folowing facts:

- Recession of galaxies
- The universal black body radiation at 2.7 K and its isotropy
- The helium abundance and homogeneity
- The deuterium abundance

All other explanations are far less satisfactory.

Fig. 2: Do you believe in big bang?

The nucleonic density ρ_N is far more uncertain. A lower limit of $\rho_N = 10^{-31} g/cm^3$ ($n_N = 10^{-7} cm^{-3}$) can be obtained from the average density of the "visible" matter in the form of galaxies ((Shapiro, 1971). A lot more matter could be present under the form of "invisible" matter such as dark stars, "black holes", intergalactic gas (ionized but not too hot for X-ray emission) low energy neutrinos, unknown types of particles or even ghosts and angels of all celestial classes. (Whether they have a rest mass or not, these various objects will have an energy associated with their motions, hence will participate in the mean gravitational field and cause some retardation on the expansion of the galaxies!) Although the deceleration of the galaxies is too small to be detected, an upper limit can be set, which also can be used to set an upper limit on the mass of "invisible" matter. A value of $\sim 3 \times 10^{-29} g/cm^3$ is arrived at, ($n_H \sim 3 \times 10^{-5}$), comparable to the so called "critical density" needed to "close" the universe (zero total energy). (See Peebles (1971) for a thorough discussion.)

In Fig. 3 the history of nucleosynthesis in the first hours is depicted graphically for a choice $\rho_N =$ $= 2.25 \times 10^{-31} g/cm^3$. As the temperature decreases, some neutrons are captured by protons to form ^2H, which can

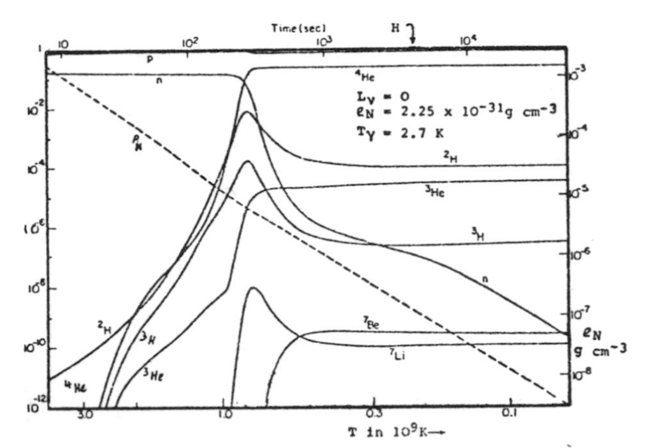

Fig.3: Evolution of nuclear abundances and baryon density (dotted line) during the expansion of a typical ($h_o=10^{-4.5}$) "standard" (ξ = C = 1) big bang model. (From Wagoner, 1973)

itself be destroyed by n or p capture into ^{3}He and then finally in ^{4}He. ^{4}He is such a stable structure that it is almost a dead-end. Only very weak branches are allowed to proceed from there into ^{7}Be (which decays into ^{7}Li) or into ^{7}Li itself. The lack of stability of mass 5 and 8 (both due to the great stability of ^{4}He) prevents the formation of important quantities of heavier nuclei.

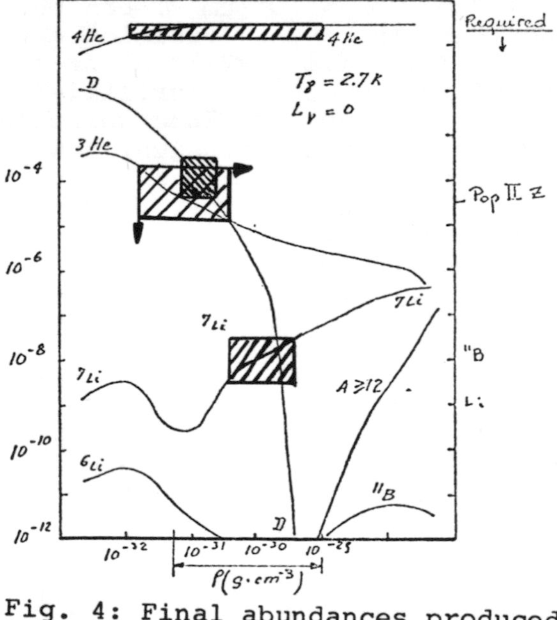

Fig. 4: Final abundances produced by standard (ξ = C = 1) big bang models, whose only parameter is h_o (or present baryon density). (From Wagoner, 1973)

In Fig. 4 the final yield of light elements is plotted as a function of the present mean universal density ρ_N (left here as a free parameter). Toward the low value of ρ_N ($\sim 10^{-33}$) the big bang yields nothing but protons. This comes from the fact that in the big bang the only way of producing heavy atoms is to go through the (p+n) capture process, hence the fate of the neutrons is crucial (see Figs.5 and 6).

The neutron/proton ratio:

As long as the reactions

$$p + e^- \rightleftarrows n + \nu$$
$$n + e^+ \rightleftarrows p + \bar{\nu}$$

have lifetimes shorter than t (universe)

$$n_p/n_n = \exp\,[M_n - M_p]\,C^2/kT$$
$$= \exp\,[1.3\,\mathrm{MeV}]/kT$$
$$\simeq 1 \text{ at high kT}$$

Around kT = 1 MeV n(positron) and n(electron) fall rapidly and n/p is no more in equilibrium.

In fact all n decay freely by $n \to p + e^- + \bar{\nu}$ $\tau \simeq 10^3$sec.

Fig.5: Nucleosynthesis in big bang.

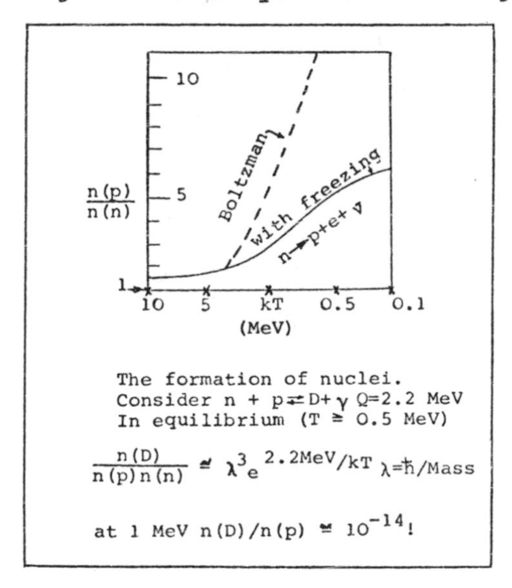

Fig.6: The n/p ratio.

The neutrons, formed in first seconds of the universe can either betadecay to protons in $\sim 10^3$sec, or capture a proton to form a deuterium (D) and start the chain. In the low ϱ_N range betadecay competes successfully with neutron capture and one would get a purely hydrogenic big bang (see Fig. 7).

Around T = 1 MeV, $\gamma + D \rightarrow n + p$ stops
What happens to the neutrons?

Either $n \rightarrow p + e + \bar{\nu}$ ($\tau_\nu \simeq 10^3$ sec)
hence no nucleosynthesis

or $n + p \rightarrow D + \gamma$

$\tau_D = \frac{1}{np} \frac{1}{\langle \sigma v \rangle}$, if $\varrho_N \simeq 2 \times 10^{-8}$ g cm^{-3} $\tau_\nu \simeq \tau_D$

Then (A) if ϱ_N(T=1MeV) $\gg 2 \times 10^{-8}$ g cm^{-3}
all the n form D and He^4 but at T = 1 MeV $n/p \simeq 0.2$ hence the ratio
$He^4/H = \frac{(2n+2p)}{p} = 0.1$!

(B) if ϱ_N (T=1MeV) $\ll 2 \times 10^{-8}$ g cm^{-3}
all the neutron decay: $\frac{He}{H} = 0$

The crucial question:

what was ϱ_N when T = 1 MeV?

Fig.7: Mechanism for He-formation in big bang.

Around 10^{-33} g/cm^3, the (n+p) process takes place for a fair fraction of the neutrons and D is made in large amount. At larger densities, D is itself destroyed into ^{3}He and ^{4}He by further nuclear capture processes. Above 10^{-32}g/cm^3 one gets more than 20 % ^{4}He by mass and above 30 % above 10^{-29}g/cm^3. This is the range of ^{4}He observed in various locations of our galaxy and in a number of neighbouring galaxies. Hence we have the rather comforting situation that within the range of universal densities allowed to us by cosmological observations big bang nucleosynthesis predicts the range of ^{4}He values obtained from totally different observations.

Figs. 3 and 4 are made under the common assumption that the universe contains an equal amount of neutrinos and antineutrinos or, in other words, that the chemical potential of the neutrinos is zero, or in other words that the leptonic number of the universe

$$L_\nu = \frac{n(e^-) - n(e^+) + n(\nu) - n(\bar{\nu})}{n(\gamma)} = \frac{n(p)}{n(\gamma)} \ll 1$$

in which case big bang predicts the existence of a black body radiation at 2.0 K with a ratio $n(\nu)/n(\gamma) = n(\bar{\nu})/n(\gamma) = 0.$

Notice incidentally that this represents a density of 40 ν and $\bar{\nu}$- per cm^3 circulating everywhere including within us! But because of their low interaction cross-sections they don't disturb us much, nor do they disturb our most sensitive detecting devices: there is just no hope of seeing them today.

But in the seconds of big-bang nucleosynthesis they did play an important role. This is best seen by varying the number L_ν or equivalently the neutrino chemical potential. This question will be considered again when we shall make comparison with the observations. For the moment we simply note that D, ^3He, ^4He and ^7Li are the only isotopes which are made in interesting amounts during the big-bang (and this will remain true even after varying L_ν). By "interesting" I mean that they have at least a chance of accounting for the observations. And we note that D is a much more sensitive universal "barometer" than ^4He since its yield varies by several orders of magnitude in the cosmologically allowed range of baryonic densities. The same is true, but to a lesser extend, of ^3He and ^7Li.

III. PREGALACTIC NUCLEOSYNTHESIS

This is a hypothetical stage preceding galactic birth, and perhaps causally related to it. The need for this stage, as far as nucleosynthesis is concerned, comes from the fact that even the oldest stars in our galaxy contain a small but definite amount of metals (elements heavier than helium: the Z of the astronomers) which they must have gotten somewhere (the big-bang appears thoroughly unable to account for these heavy atoms). Hence the idea that before the birth of the galaxy some form of nucleosynthesis must have taken place (Truran and Cameron, 1971).

Pregalactic events have also been invoked as an alternate way of accounting for the ^4He abundance (in case the big-bang model would, in the future, meet with insuperable difficulties) (Wagoner et al., 1967).

It has been customary to assume that the pregalactic "events" involved Super Massive Objects (SMO), short-lived and violently-exploded. An extensive study of these so-called little bangs has been made by Wagoner. Each object can be described by essentially three free parameters, which are related to its mass, the initial total neutron to proton ratio and the strength of the explosion. Interesting amounts of the light elements D, He^3 and Li^7

and of the metals could be obtained (together with plenty of ^4He) with choices of parameters which are highly unrealistic, or hardly defendable, except in a very "ad hoc" fashion (Wagoner,1973). Since then, Fricke (1973) has studied in details the physics of these objects and found that the implosion-explosion feature is not realized for metal-free initial composition.

Two other difficulties appear when comparisons are made with observations: the close universality of the $^4He/H$ ratio is hard to reconcile with anything except a thoroughly universal mechanism (like big bang); little bangs are "local" sources and, unless something special in the physics of the problem requires it, it is hard to see why they should always result (after dilution with unprocessed matter) in the same He/H ratio. Also, as shown by Wagoner et al. (1967) and Audouze and Fricke (1973) small bangs will function on the hot CNO cycle, which always results in large ratios of C^{13}/C^{12}, N^{15}/N^{14}, O^{17}/O^{16} etc.. Observations of some halo old stars (Cohen and Gradsdalen, 1968) show that C^{13}/C^{12} ratios are rather normal (< 0.10). We may then argue that not more than 10^{-2} of the galactic mass can have been processed by such objects; their effects would be seen in these isotopic ratios of the present cosmic abundances.

For these reasons, the SMO hypothesis is presently in difficulty. Other forms of pregalactic nucleosynthesis may, nevertheless have taken place. We may fancy that generations of galaxies have existed before our own and may have contaminated the intergalactic gas, just as stars are believed to contaminate the interstellar gas; the age of the universe is not so well known, after all, as to raise difficulties with that hypothesis.

Until these problems are understood, it is reasonable to consider pregalactic nucleosynthesis as not yet establish and to forget it in our discussion.

IV. NUCLEOSYNTHESIS IN THE GALAXY

Nucleosynthesis during the galactic life can be classified in three stages, according to the energy of the element-forming nuclear interactions: low, intermediate and high energy stages.

(a) The stellar thermonuclear phase

The origin of heavier elements (from carbon to uranium) is generally attributed to thermonuclear reactions in stellar interiors (Burbidge et al., 1957) (where the operating energies are from tens to hundreds of KeV).

The same reactions can, in principle, generate some of the light elements. The most important one is of course ^4He but ^3He and ^7Li are also candidates. Although the formation of deuterium (by $p + p \rightarrow D$) is one fundamental step in stellar energy generation, its destruction rate (through $p + D \rightarrow {}^3He$) is so much faster than its formation rate that its equilibrium abundance is entirely uninteresting in our context. The other isotopes (Li^6, Be^9, B^{10}, B^{11}) are bypassed by the normal chain of thermonuclear reactions and are not produced in stars. Furthermore, even if they were produced there, they would not survive the internal heat: in typical stellar lifetimes (10^8 - 10^{10}y) D is destroyed above 0.5×10^6 K, Li above 2×10^6, Be above 4×10^6 K and B above 5×10^6K; such temperatures are reached in the largest fraction of Main-Sequence stars except of course in the outer layers and the photosphere. However, stars later than F have surface convective zones which are expected to carry the surface layers to inner (hotter) layers. For instance, we are not surprised to find a notable absence of deuterium in the solar photosphere and in the solar wind: the original solar D has all be transformed in ^3He long ago. In the same fashion the solar Li is about 10^2 smaller than the meteoritic Li or stellar Li as found in very young stars. This point will be discussed again later on.

In later stellar phases (Red Giant), stars develop much stronger and deeper surface convective zones, which probably achieve to destroy even the surface content of these light isotopes (the special status of ^3He, ^4He and ^7Li will now be discussed separately).

The process of energy generation in Main-Sequence stars involves the transformation of protons in ^4He. An appreciable fraction of these atoms is likely to emerge unscathed from further stellar evolution, and to return to interstellar space at the end of the life of the star. However, as discussed later, the contribution of this stellar-formed ^4He to the "cosmic" ^4He is most likely small.

The hydrogen-burning phase generates a small amount of ^3He: typically, in the ashes of hydrogen, ratios of

$^3He/^4He$ of 10^{-2} to 10^{-4} are reached (Iben,1969). We do not know what fraction (if any) of this ^3He will be preserved from further burning, and return to space.

The yield of ^7Li in "normal" thermonuclear phases is negligible: Main-Sequence stars slowly destroy their initial lithium abundance. However, the detection of large Li abundance in some types of Red Giants suggests that this isotope is somehow generated during this late stage. It has been suggested by Cameron and Fowler (1971) that some type of thermal pulses, in shell-burning layers, may generate some ^7Li through $^3He + {}^4He \rightarrow {}^7Be \rightarrow {}^7Li$. This hypothesis has been considered in detail by Ulrich and Scalo (1972) and also by Sackman et al. (1972). The subject contains many poorly-known parameters and remains unclear.

At any rate these atoms of ^7Li can only contribute to the cosmic abundance of Li if, through stellar winds, they are massively returned to space (Reeves et al.,1973). Again we have, as yet, no quantitative estimate of the importance of this effect.

(b) Shock wave nucleosynthesis in supernova

Although little is known about the mechanism of a supernova (SN) explosion, it is plausible that, somehow, the passage of a shock wave through a thin stellar envelope is involved. As it reaches the outermost layers, the wave reaches velocities corresponding to several MeV per nucleon. Colgate (1973) and Hoyle and Fowler (1973) have considered the possibility that, in view of these high velocities, the He atoms of the atmosphere may be destroyed (spalled) into neutrons and protons which, after the passage of the shock, may recombine into deuterons. In the same fashion, nitrogen atoms can be spalled in boron-11.

The main problem is a problem of energetics which can be separated in two questions; what is the total energy available in the exploding objects and what is the fractional mass of the object in which high enough velocities will be reached (25 - 30 MeV for the spallation of ^4He and 7 to 10 MeV for the spallation of nitrogen).

To the first question we can already give some meaningful answers from recent observation of supernovae (Kirshner et al.,1973) and of supernovae remnants (Lecce conference on supernovae 1973). Typical energies of type II supernovae are $1\text{-}2 \times 10^{51}$erg in an ejected mass of $\sim 5\ M_\odot$ corresponding to 0.2 MeV per nucleon.

To the second question we have only simplified models due mostly to Colgate. Using the numbers given by him in a recent paper, and the energy values given above makes it clear that the amount of D and ^{11}B generated in supernovae falls short, by a rather large factor, of meeting the observational abundance requirements (Reeves, 1973).

One may also consider supermassive exploding objects (SMO) at the beginning or during the life of our galaxy. There again we can use the low values of isotopic ratios such as $^{13}C/^{12}C$ etc. to put a limit on the galactic mass which has been processed this way (at most one percent). There again, consequently, we come out with uninteresting yield.

The efficiency of shock-wave acceleration may be better than estimated by Colgate. The observations of Kirshner make this unlikely: the mass distribution at high velocity is rather small. In view of these difficulties it appears highly improbable that shock wave nucleosynthesis contributes importantly to deuterium and boron-11 nucleosynthesis.

(c) High-energy phase: Galactic Cosmic Rays (G.C.R.)

Fast particles have been pervading the interstellar medium for a very long time. In Fig. 8 their energy spectrum both in the vicinity of the earth and in interstellar space is shown. The interstellar space spectrum is obtained

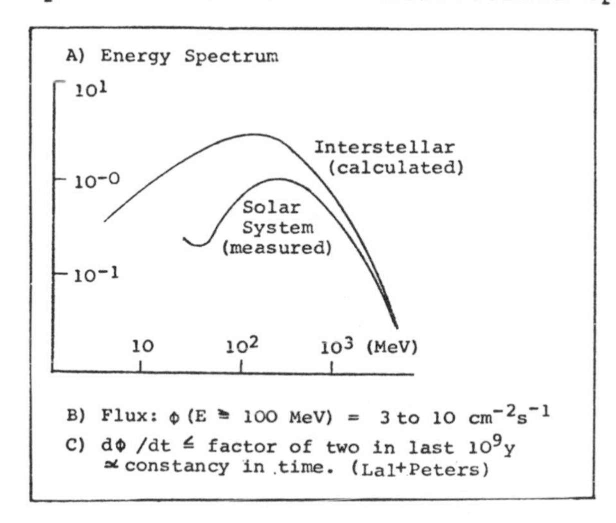

Fig.8: Facts of life of Galactic Cosmic Rays (G.C.R.)

after computation of the flux reduction brought by the repulsion effect ("modulation") of the solar wind on the incoming flux.

After more than fifty years of studies, we still do not know where the galactic cosmic rays come from. We do not even know if the bulk of them is of galactic origin or if they pervade the entire universe, although the preliminary results on the very-heavy nuclei does seem to indicate a galactic origin: indeed the ratio of the actinides (Z = 90 to 100 where most of the nuclei are unstable to fission) to the lead region (Z = 80 to 82) would appear to indicate (Schramm,1972) a rather local origin (at least within our cluster of galaxies). For reasons of energetics, a local origin would also be indicated, although there we may still have some surprises.

The chemical composition of galactic cosmic rays is shown in Fig. 9. There appears to be an overabundance of heavy elements with respect to "cosmic" abundances in the source of cosmic rays.

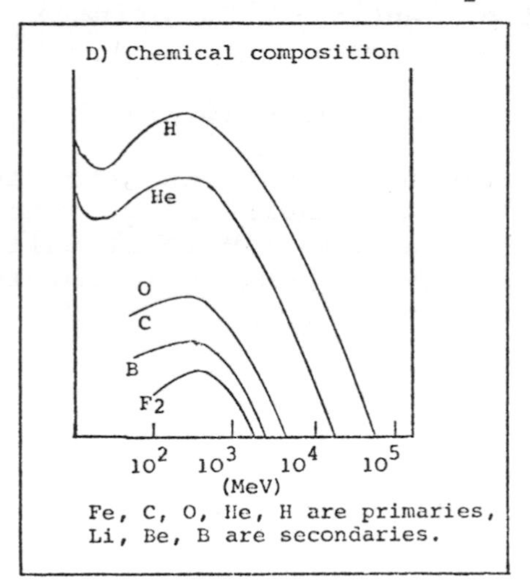

Fig.9: Nucleosynthesis in Galactic Cosmic Rays.

For many years this was taken as a proof that galactic cosmic rays originate in supernovae since SN explosions would provide both the energy and the chemical enrichment (Ginzburg,1964). However, in recent years, chemical over-abundances of heavy elements have also appeared in the solar flares (where the raw material is "normal") pointing out to the fact that the over-abundances may reflect the selectivity of the acceleration processes (whatever they are) (Casse and Goret,1973). There, again, the very heavy cosmic rays may help: typical explosive processes would generate a peak distribution around Pt (by rapid neutron nucleosynthesis). The ratio of Pb to Pt,for instance,would reflect the occurrence of such an event. Preliminary investigations (Schramm,1972) would confirm the presence of a Pt peak but the situation is still very uncertain. (Needless to say, if the peak is absent and if G.C.R. are just ordinary matter with some funny preferential acceleration then the amount of actinides would be much smaller and would depend on the age of this matter. The previous test on the non-universality of the G.C.R. would probably become meaningless. All these things are interwoven, and it is not easy to disentangle them.)

One interesting feature of these chemical abundance ratio is the large abundance of the rare elements Li, Be, B and also D, He^3 as compared with natural abundances. In Fig. 9 the B abundance, for instance,is larger than the Fe abundance, while on the sun it is several thousand times smaller. The reason for this is that B nuclei are formed through collisions of the fast CNO on the interstellar atoms of H and He (mainly). This process is called spallation in the jargon of nuclear physicists (Bradt and Peters,1950).

The ratio of Li Be B /CNO in G.C.R. ($\sim$ 0.20) reflects the amount of matter traversed by these fast particles from their sources to us. One finds approximately 7g per cm^2 of matter (assuming a homogeneous source region). If the G.C.R. were confined to the region where spallation takes place (presumably the galactic disk) the ratio would go up to 70 % (there would be a balance between nuclear production and nuclear destruction). The fact that the ratio is 0.20 and not 0.70 shows that the G.C.R. are not confined to the galactic disk.

What happens to these fast ions? Three possible fates: either they escape from the galaxy or they are destroyed by spallation reactions on the interstellar atoms or they are decelerated by electron collisions and they mix themselves with the interstellar gas (see Fig.10).

Is it possible that this process could have contributed importantly to the "cosmic abundance" of any element or isotope? Since this process is very slow, we look at some of the rarest isotopes as being the only ones for which this origin could be important. Let us look for instance at the light isotopes D, ^3He, Li, Be, B and start

(A) Fast CNO on p,α at rest
CNO ⟹ p,α ⟹ Li, Be, B
High recoil velcocity!

What happens to them?

I. Escape from galaxy
II. Undergo nuclear destruction
III. Decelerate to rest by electronic collisions and enrich the gas.
We care only about III.

(B) Fast p,α on CNO ... at rest.
Li, Be, B are generated essentially at rest.

Fig.10: Production of light nuclei by spallation of interstellar gas.

with the case of ^{9}Be. Its typical stellar abundance is N(Be)/n(H) ~ 2 x 10^{-11}. The rate of formation by G.C.R is given by (Reeves et al., 1970) (in rather simplified way):

$$\frac{d\ n(^9Be)/n(H)}{dt} = \phi_{p.\alpha}\ (E \geq 30\ MeV)\ \frac{n(CNO)}{n(H)}$$

$$\sigma(CNO + p.\alpha \rightarrow {}^9Be)$$

where $\phi p._\alpha$ (E ≐ 30 MeV) is the flux of G.C.R. of p and α with enough energy to break the most abundant targets (C.N.O.) into ^{9}Be (~ 10 $cm^{-2}sec^{-1}$), $\sigma[(CNO+p.\alpha) \rightarrow {}^9Be]$ is an average value of the cross section (~ 5mb = 5x $10^{-27}cm^{-2}$) and n(CNO)/n(H) ~10^{-3} is the abundance of these targets in interstellar matter. This way we get d[n(^{9}Be)/n(H)] / dt ≈ 5 x $10^{-29}s^{-1}$ and the time required to build up typical stellar abundances is t ~1.3 x 10^{10}y which is quite similar to the age of the galaxy. This approximate agreement shows that the following picture is tenable: the atoms Be are gradually produced in interstellar space by the effect of the G.C.R., and when a star is formed out of the galactic gas, it contains these isotopes which become visible in the stellar photosphere.

If we want to know what other elements are made this way, and, in particular, if we want to compare the isotopic and element ratios, we have to make a complete model in which we take into account in detail the various physical factors of a propagation-diffusion model of G.C.R. in the galaxy.

From the results of this computations (Meneguzzi et al., 1971) we may obtain some element and isotopic ratios. They are given in the following table.

D/H	$^3He/H$	$^6Li/H$	$^7Li/H$	$^9Be/H$	$^{10}B/H$	$^{11}B/H$
4×10^{-8}	4×10^{-8}	10^{-10}	2×10^{-10}	2×10^{-11}	10^{-10}	3×10^{-10}
$^3He/^4He$		$^7Li/^6Li$			$^{11}B/^{10}B$	
4×10^{-7}		~ 2			~ 3	

Since the time integrated G.C.R. flux is not known (G.C.R. were possibly stronger in the early days of the galactic life), the solar-meteoritic value of Be is taken as a normalizing value. Actually we get just this value if we assume a constant G.C.R. rate for 10^{10}y.

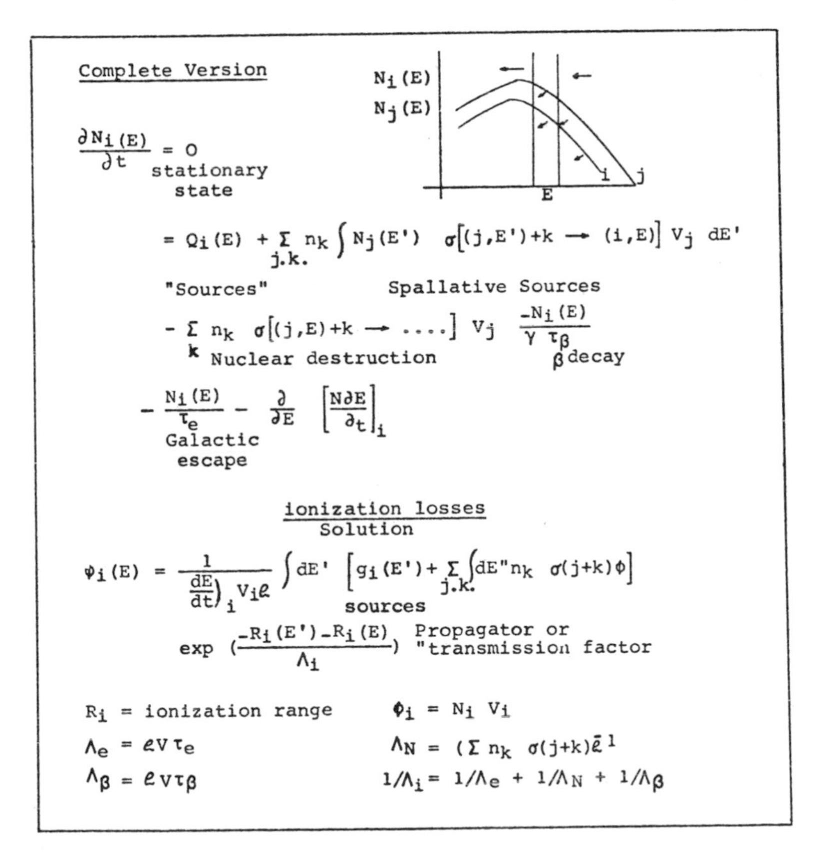

Fig.11: The structure of the calculation of the formation and deceleration of light elements in interstellar space.

V. THE EFFECT OF STELLAR EVOLUTION ON NUCLEOSYNTHESIS OF THE LIGHT ELEMENTS

Before we can fruitfully compare the formation rate to the observed abundances we have to take into consideration the effects of stellar processes on these atoms. We call "astration" the process by which a mass of interstellar gas is first contracted into a star, and later returned into space. One important effect of astration is of course the formation of heavy elements from light atoms (H $\rightarrow$ He; He $\rightarrow$ C,O; C,O $\rightarrow$ Mg, Si etc.) but also the destruction of some fragile nuclei (Li Be B $\rightarrow$ He). As discussed before, it is probably not a bad approximation to state that when a mass of gas goes into a star and is later ejected (i.e. when a mass of gas is astrated) it looses its whole content of D, Li, Be, B (for ^{3}He the situation is not clear) and unless some special processes manage to recreate some of these atoms in the very late stages of the stellar life, astration is equivalent to destruction. Let us for instance consider the case of deuterium which is formed in the big bang. The mass fraction of the original deuterium which will have survived in our galaxy is, according to these views, simply given by the mass fraction originating in the big bang times the mass fraction of the interstellar gas (f) which has never (not even once) been astrated.

It is not easy to obtain the value of the "unastrated" mass fraction of the gas. Some estimates can be obtained by models of galactic nucleosynthesis taking into account stellar luminosity functions, stellar formation rates and observations of mass ejection from stars. It would seem that a value of $f \sim 0.2$ should be reasonable (obtained from the models of Salpeter,1955; Schmidt 1963; Van den Bergh,1957; Truran and Cameron,1971). Hence we shall expect the fragile products of big bang nucleosynthesis (D, ^{3}He, ^{7}Li) to have decreased to about one fifth of their original value today (and to about one third at the birth of the sun: $f(t_\odot) \sim 0.3$)). Astration will also influence the abundance of the elements Li,Be,B, this time in three different ways. First, since the galactic cosmic rays (G.C.R.) are believed to originate from supernovae(SN), the flux of G.C.R. must be related to the rate of SN activity, and hence to the rate of stellar activity which, from arguments of nucleosynthesis, was probably appreciably stronger in the early days of the galaxy (Schwarzschild,1958).Secondly,since the interstellar targets (CNO nuclei) are themselves issued from astration we may expect their abundances to have increased with time from very small values to their present interstellar values. Thirdly, after their formation, the L elements (Li, Be, B) will also be partly destroyed through further astration. The first two effects more or less compensate each other (more cosmic rays; less targets). As a first approximation we may think that, in the first few billions of years the increased rate of stellar activity was paralleled by increase in the astration rate, so that whatever happened in these early days did not contribute significantly to the (time-integrated) abundances of today. The major contribution comes mostly from the quieter days extending all the way from this early period till today ($\sim 10^{10}$y). Hence the simple approximation of a constant G.C.R. (as in Meneguzzi et al., 1971) is probably not bad.

VI. SUMMARY OF THE OBSERVATIONS: THE EVOLUTIONARY ABUNDANCE CURVES.

In the following pages the observations relevant to each isotope are discussed and the EAC are tentatively drawn from the scanty data combined with the theoretical models of nucleosynthesis described in these notes. Much more data will be necessary before we can definitely select the appropriate evolution factors.

(a) Deuterium

On D atoms themselves we only have upper limits: Weinreb (1962) through negative observations of the 92 cm spin-flip line of D (the equivalent of the H21 cm line) givesa limit $n(D)/n(H) \leq 7 \times 10^{-5}$ in front of Cas A. Recently Cesarsky et al. (1973) have observed this same line in front of the galactic center and gave

$$3 \times 10^{-5} < n(D)/n(H) < 4 \times 10^{-4}$$

Their lower limit is actually rather weakly established and the value could be much smaller (Pasachoff, private communication).

Another upper limit comes from solar wind observations of $^3He/^4He$ ratios, (Geiss and Reeves, 1972), if one notices that the present day ^3He abundance in the solar surface is actually the sum of the protosolar D and ^3He: the protosolar D has been burned into ^3He in the early solar days. The solar wind ratio of $n(^3He)/n(^4He) = 4 \times 10^{-4}$ (recently roughly confirmed by coronal measurements (Hall, 1972)) gives $\frac{n(D) + n(^3He)}{n(H)} = 4 \times 10^{-5}$ at $t_\odot$

(the birth of the sun) and hence $n(D)/n(H) \leq 4x10^{-5}$ at that moment.

Other measurements of D exist on various molecules: HDO (earth and meteorites) (Boato,1954) CH_3D (Jupiter) (Beer et al., 1972) HDN (Orion Nebula)(Jefferts et al.,1973) and recently in HD in a number of relatively dense clouds (Morton et al., 1973). For all these cases we have to worry about D enrichment by molecular exchanges between D and H as for instance in

$$H_2 + D \rightleftharpoons HD + H \quad \text{and}$$
$$HD + H_2O \rightleftharpoons HDO + H_2$$

Because of its larger mass, the zero-point energy of the vibrational spectrum of the molecule containing D is always at lower energy than that of the equivalent molecule containing H. Hence, at low temperature, the D atoms will migrate toward the molecules more readily than H atoms. The consequent D enrichment can be computed from knowledge of the partition functions of the molecules involved, and can be quite high in the low T range typical of interstellar clouds. For instance in the Orion nebula (at $T \sim 100$) the large $DCN/HCN \sim 3x10^{-3}$ observed by Jefferts et al.(1973) can well be accounted for by a $n(D)/n(H) \sim 3x10^{-5}$ as discussed before (Solomon and Woolf, 1973; Reeves, 1972).

The D/H ratio in water ($1.6x10^{-4}$ on earth and meteorites) can be used, together with an evaluation of the formation temperature of the meteorites ($\sim$ 300 K) to obtain a value of n(D)/n(H), through the reactions (Geiss and Reeves ,1972) in stars.

$$H_2O + HD \rightleftharpoons HDO + H_2$$

Again a value of $n(D)/n(H) \sim 3\text{-}5x10^{-5}$ is obtained. Finally the CH^3D/CH^4 ratio recently obtained in Jupiter (Beer et al.,1972), coupled with a very uncertain estimation of the formation temperature of CH^3D in the Jupiter's atmosphere yields n(D)/n(H) between 3 and 7×10^{-5}.

Trauger et al. (1973) have recently reported a measure of HD/H_2 in Jupiter's atmosphere. Since the Jovian atmospheric hydrogen is likely to be largely molecular, their measurement ($D/H = 1.5x10^5$ within a factor of two) is probably a fair estimate of the protosolar deuterium abundance.

Putting all this information together, the data is coherent with the following statement

a) the cosmic n(D)/n(H) is $2.5\pm1\times10^{-5}$ (today and also at the birth of the sun)

b) higher values obtained in molecules are due to enrichment by various effects including molecular exchange reactions.

Correcting for astration and converting into mass fraction X(D) these numbers correspond to a pregalactic X(D) of 5×10^{-5} to 2×10^{-4} which can be obtained in the big bang for present universal baryonic density of $2\text{-}3\times10^{-31}$, a few times larger than the density of visible matter but much below the "critical" density. The EAC of D is traced, according to these observations and models in Fig. 12.

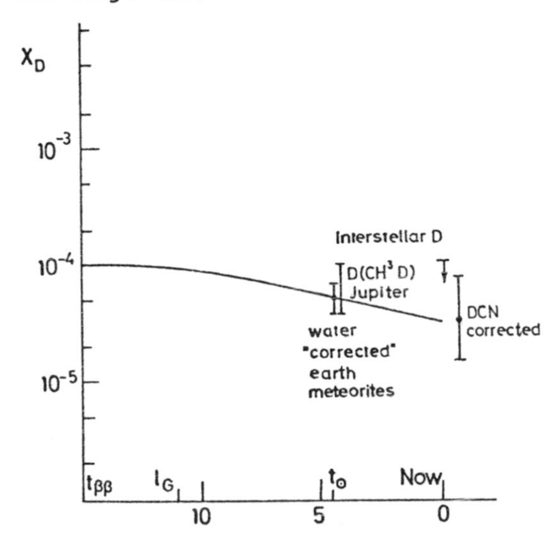

Fig.12: The Evolutionary Abundance Curve (EAC) of D. X(D) is the fractional amount of D. The point at t_{BB} is the result of big bang nucleosynthesis for a present universe density and microwave background temperature $\rho_b \simeq 3 \times 10^{-31}$; $T = 2.7^{\circ}K$, and leptonic number $L_e = 0$. Deuterium depletion in the pregalactic era is assumed to be small: it becomes significant at the birth of the galaxy t_G and continues till now. The observational values and limits (observations of the deuterium abundance at the formation of the solar system, and in the interstellar medium) are discussed in the text.

(b) Helium-3.

It is very difficult to obtain a correct evaluation of the cosmic abundance of ^3He, since ^3He can be locally enriched either by spallation reactions(in some meteorites the ratio $^3He/^4He$ reaches the value of 0.1),or by thermonuclear burning of D. As discussed before, the present ratio of $^3He/^4He$ in the sun or in the solar wind is $\sim$ 4×10^{-4} corresponding to $^3He/H \sim 4 \times 10^{-5}$ and this is the sum of the protosolar D and ^3He. Since we have estimated D to be D/H: $2.5 \pm 1 \times 10^{-5}$ we get $^3He/H \sim 1.5 + 1 \times 10^{-5}$ (protosolar).

This value is not in disagreement with the number given by Black (1971) from a study of certain classes of meteorites which have solar wind implanted He atoms with the ratio $^3He/^4He \sim 10^{-4}$. Black argues that these atoms came out of the sun before D-burning took place, and hence should represent the protosolar $^3He/^4He$ ratio (but diffusion could also have played a role).

Contrary to D, ^3He can be obtained by stellar evolution: in the ashes of an H burning region the ratio of $^3He/^4He$ will take values from 10^{-4} to 10^{-2}. If these ashes succed in escaping the star before further temperature increase (for instance as stellar wind in the Red Giant phase) they could contribute appreciably to the cosmic abundance of ^3He. Let us for the moment assume that this is not the case and that, as for D, the bulk of ^3He comes from the big bang.

If we choose f = 0.3 at the birth of the sun (as for D) we get $X(^3He) = 4 \times 10^{-5}$ to 1.5×10^{-4} corresponding to densities of 1 to 2×10^{-31} g cm^{-3}. Since the universal density could not be much lower than this, we are left with very little space for formation of ^3He in stars. Of course f could be much smaller than 0.3, in which case one would need to invoke a lot of stellar He (but we would get into trouble with D).

The only conclusion we can reach is that with a density of 2×10^{-31} we can account for D and ^3He without invoking any stellar formed ^3He. The EAC of ^3He is shown in Fig. 13.

(c) Helium-4.

^4He abundances have been measured or estimated in very many places and, in the large majority of cases,the results are consistent with $^4He/H = 0.1 \pm 0.03$ (or 20 to

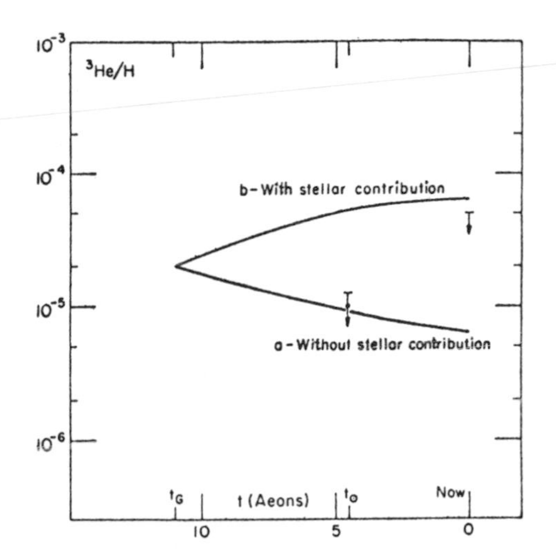

Fig.13: The Evolutionary Abundance Curve (EAC) of ^3He. The curve a) represents the case in which ^3He is destroyed in all material ejected from stars; the curve b) represents the case in which ^3He is assumed to survive in the region external to the helium zone in low mass stars. The upper limits pertain to the protosolar gas and to the (present) interstellar gas (From Reeves et al., 1973).

35 % in mass) in our galaxy as well as in a number of neighbouring galaxies (Danziger 1970 ; Searle and Sargent, 1972). The homogeneity of this ratio over such a wide span of space strongly indicates that a "universal" process has been at work and the best candidate is of course big bang. The extent at which astration has altered the original He/H is generally recognized to be small (from galactic evolution model). Attempts have been made at finding variations of the He/H from which one could separate the big bang contributions (at a density of 2×10^{-31} the yield of ^4He is about 22 % (or $^4He/H \sim 0.08$)) from addition due to stellar nucleosynthesis, by study of radio recombination lines in H11 regions (Mezger, 1973). A few points need to be clarified for the interpretation to become fully convincing. The EAC, with and without stellar contribution, is presented in Fig. 14.

(d) Lithium-6 and Lithium-7.

Lithium has been observed in a large number of stellar objects (Zappala, 1972 ; Herbig and Woolf, 1966). Most of the observations can be summarized by saying that the

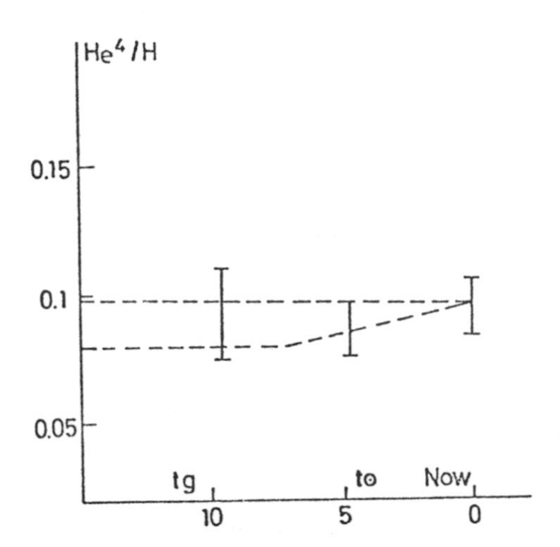

Fig.14: The EAC of ^{4}He. The upper curve assumes a negligible contribution of stellar-made ^{4}He. The lower curve assumes a rather large contribution.

abundances are generally larger in young stars than in old stars for stars of a given spectral class, and are also larger in early-type stars than in later-type stars for stars of a given age (for instance in a cluster). Furthermore, the higher abundances found in very young stars or very early stars are in agreement with the values of T Tauri stars (the very first stage of stellar evolution and in the meteorites (when properly normalized to Si abundance) one gets $n(Li)/n(H) \sim 1$ to 2×10^{-9}. The standard picture is that this should be the initial value inherited from the interstellar gas and should be the same for all stars. Then, as the star is formed, it gradually burns its Li content through thermonuclear reactions partly during the pre-main sequence evolution (at least for late-type stars) but mostly during the main-sequence evolution. Stars later than F have a surface convective zone which brings the surface material to higher temperatur at the bottom of this zone. The later the spectral type, the hotter is this temperature. As a result the light element may be gradually depleted from the surface and the depletion increases both with age and with spectral type when we compare two stars of the same age (Herbig and Woolf,1966; Wallerstein and Conti,1969).

This picture has much success in reproducing qualitatively the observational trends. Quantitatively, things are not so satisfactory. For instance, the depletion of solar lithium should not be so strong, if computations are made according to standard ideas (Ezer and Cameron, 1965). There is a possibility that other effects are at work such as diffusion of lithium downward from the solar convective zone through the combined action of gravity and light pressure (Michaud and Vauclair, in preparation).

The ratio of $^7Li/^6Li$ in meteorites is 12.5; in stars we have only the result $^7Li/^6Li > 10$. Combining these data a consistent value of $(^6Li)/H$ in protostellar matter is 1 to 2 10^{-10}.

There is no doubt that the ratio of $^7Li/^6Li = 12$ cannot be reproduced in the standard picture of galactic cosmic rays (unless a strong flux of low-energy particles is arbitrarily added).

Furthermore, while the ratio $^6Li/H \sim 10^{-10}$ seems to be coherent with the prediction of the G.C.R. calculation, the value $^7Li/H \sim 10^{-9}$ is probably too high. Hence it is reasonable to suggest that ^6Li is an authentic child of the G.C.R. but the paternity of ^7Li must be sought elsewhere. We have seen that ^7Li can also be produced by big bang but after correction for astration the value required is $X_{Li} \sim 10^{-8}$ which will not be reached in the big bang unless present densities $\rho_N \simeq 10^{-30}g/cm^3$ are chosen (Wagoner, 1973). At densities compatible with D, ^3He, ^4He production ($\sim 2 \times 10^{-31}$) the yield of Li is more than ten times too small. This situation can be remedied if instead of assuming (as is usually done) that the universe contains an equal number of neutrinos and antineutrinos $n(\nu)-n(\bar{\nu}) = 0$ (in the standard big bang this leads to $n(\nu) = n(\bar{\nu}) \approx 44$ cm^{-3} today at temperature 2^oK), we allow this ratio to change $[n(\nu)-n(\bar{\nu})]/n(\gamma)$ = an arbitrary number. One finds that for $[n(\nu)-n(\bar{\nu})]/n(\gamma) = -0.2$ corresponding to $n(\nu) = 28$ cm^{-3} and $n(\bar{\nu}) = 100$ cm^{-3}, one gets a big bang nucleosynthesis which gives satisfactory abundances of D, ^3He, ^4He and ^7Li at $\rho_N = 10^{-30} g\ cm^{-3}$ (today). Whether such an assumption is reasonable or not remains to be seen (Reeves, 1972). As discussed before, ^7Li could also be made in late stages of stellar evolution (giant stars) by thermal flashes.

Thus, by and large, the origin of ^7Li remains mysterious: we have three possibilities but no certitude that anyone of them works.

The corresponding EAC are shown in Figs. 15 and 16.

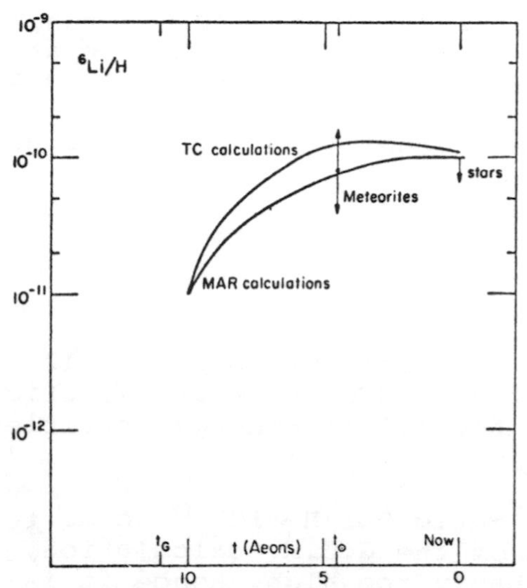

Fig.15: The Evolutionary Abundance Curve (EAC) of ^{6}Li. The curve MAR represents the (linear) growth of ^{6}Li obtained from the present rate of ^{6}Li production by galactic cosmic rays multiplied by the elapsed time. The curve TC is obtained when time varying factors (such as the GCR flux, or stellar destruction) are taken into account. (From Reeves et al., 1973)

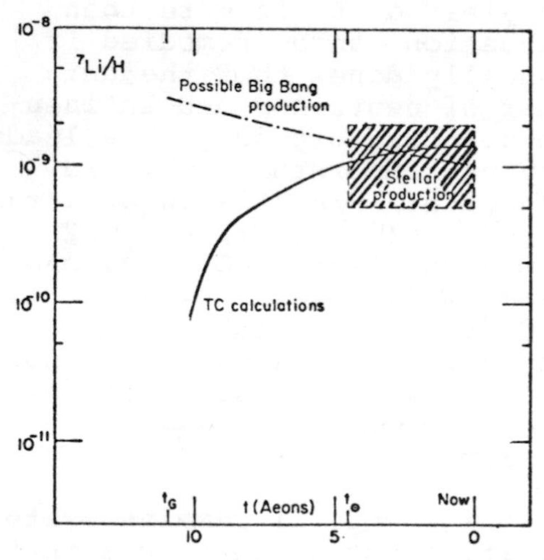

Fig.16: The Evolutionary Abundance Curve (EAC) of ^{7}Li. The curve TC represents the hypothetical effect of ^{7}Li ejected from giant stars in the interstellar medium. The dash-dot curve represents the evolutionary curve of big bang contribution if the big bang yield is as large as required (see text). The box "stellar" represents the uncertainties in the initial values for F stars. (From Reeves et al., 1973)

(e) Beryllium-9

As discussed before, the solar, the meteoritic and the average stellar value agree to give $n(^4Be)/n(H) \simeq 1$ to 2×10^{-11} (Grevesse, 1968 ; Sill and Willis, 1962 ; Merchant, 1966) . The span of stellar values extend from $< 10^{-12}$ to $\sim 4 \times 10^{-11}$ and these values appear to be correlated with stellar types in the sense that G stars have usually higher Be than F stars (Wallerstein and Conti, 1969) Since F stars have a shorter life time than G stars, they are expected, on the average, to be younger than G stars. The number of stars in this sample is very small and may not be statistically significant but if this expectation is correct the implication would be that the Be abundance in the galactic gas has been decreasing with time in the last five billion years or so. But other effects could also have been at work, related again to diffusion and the matter remains unclear.

Forgetting this correlation, the G.C.R. spallation of interstellar matter provides an adequate explanation for the mean value of Be in stars and in the solar system. The EAC is shown in Fig. 17.

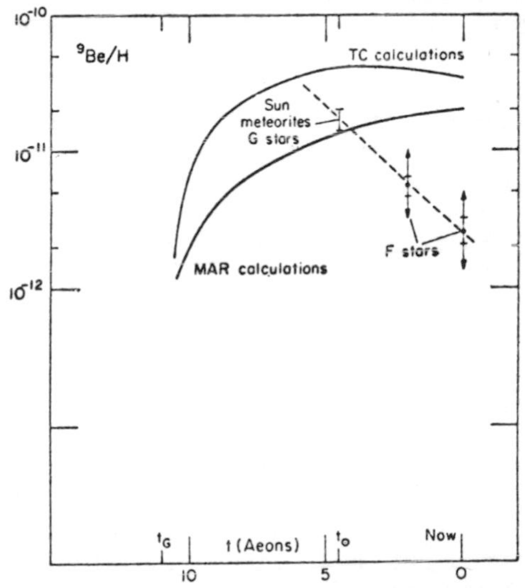

Fig.17: The Evolutionary Abundance Curve (EAC) of ^{9}Be according to Meneguzzi et al.,1971 (constant rate of production) and Truran and Cameron, 1971 (time varying rate of production). The dashed line represents the observed variation of mean Be abundance with stellar type, possibly reflecting a Be decrease with time. If this effect is real neither Meneguzzi et al. nor Truran and Cameron can account for it. (From Reeves et al., 1973)

(f) Boron-10 and Boron-11

Although we have accurate data on the isotopic ratio $^{11}B/^{10}B \simeq 4 \pm 0.4$ in the solar system (earth-moon-meteorites) the determination of the abundance of B remains highly problematic. The meteoritic value varies a great deal with the type of meteorites under consideration (Baedecker, 1971). In enstatites the abundance corresponds to $n(B)/n(H) \sim 5 \times 10^{-10}$ while in carbonaceous chondrites of type I it goes up to 3×10^{-9}. Following the generally accepted ideas that these last meteorites represent a good estimate of the cosmic abundances and, using an earlier attempt of observing the interstellar boron, Cameron selects a value of $n(B)/n(H) \simeq 10^{-8}$ (Cameron et al., 1973).

Difficulties are found, however, when other data is taken into consideration. For instance such a high value would have been observed in the sun if it was present there. Instead we only have an upper limit of $< 6 \times 10^{-10}$ (Grevesse, 1968). One may ask whether the original boron in the sun may not have been depleted by thermonuclear reactions just as Li. But if we remember that the solar and meteoritic Be agree quite closely, and since, at all temperature, Be is destroyed faster than boron, it appears very unlikely that boron could have been depleted (Morton et al., 1973)!

Furthermore, recent data on interstellar matter suggest that the high abundance of B could have been seen there and in fact has not been seen. From the limit of $n(B^+)/n(H) < 10^{-9}$ through arguments of atomic physics, a reasonable upper limit of $n(B)/n(H) < 2 \times 10^{-9}$ can be obtained (Audouze et al., 1973).

These data make it doubtful that the high abundance of C1 carbonaceous chondrites really represents the cosmic value. We may wonder whether B has not been enriched in these meteorites, just as in the case of mercury (Hg) where the upper limit of the solar abundance corresponds to a value which is about 100 lower than the C1 carbonaceous chondrites abundance. Both Hg and boron compounds are chemically known for their high mobility. In the case of mercury both the systematical trend of r and s elements and the upper limit of the solar abundance suggest that the value characteristics of enstatites should give a better estimation of the cosmic abundance. For boron a similar situation takes place since the enstatite value is coherent with the upper limit on the solar photosphere and interstellar space. One is certainly tempted to choose this value for the cosmic abundance of B. Before more information is available, however, it is wise to leave the situation open and to discuss the two alternatives (Audouze et al., 1973).

A-B/H = 5×10^{-10}, from enstatites. Then both isotopes are genuine childs of the G.C.R. where the predicted isotopic ratio (~3) is close enough to the observed ratio (~4) and everything is fine.

B-B/H = 10^{-8} from C1 chondrites. The mechanism of origin is not easily found. Hoyle and Fowler (1973) and Colgate (1973) have considered the mechanism of shock wave nucleosynthesis in supernovae envelope as a possible source of D and B. New data on SN explosion (Lecce Conference, 1973) shows that the amount of energy released is far too low and the yield D and B are orders of magnitude too small to help. Audouze and Truran (1973) and Meneguzzi and Reeves (1973) have considered high fluxes of low energy particles. Rather specialized assumptions are required in order to fit the data. In Figs. 18 and 19 the EAC of ^{10}B and ^{11}B are shown, assuming that alternative A is correct.

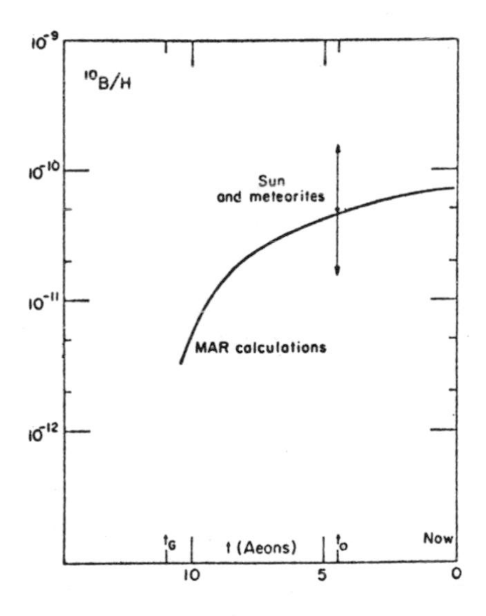

Fig.18: The Evolutionary Abundance Curve (EAC) of ^{10}B. The curve MAR is the result of constant rate of production by galactic cosmic rays (linear increase with time). (From Reeves et al., 1973)

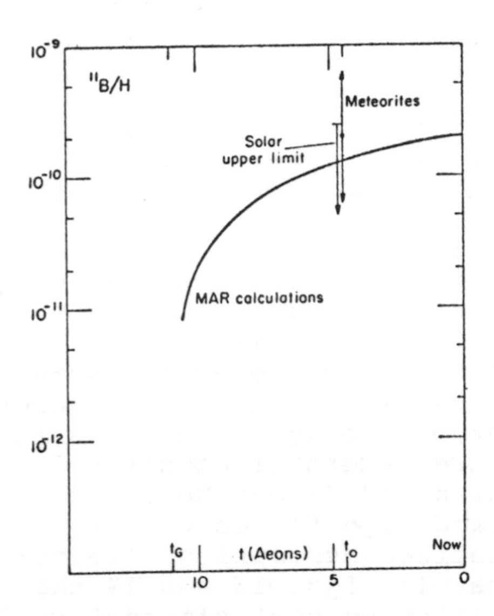

Fig.19: The Evolutionary Abundance Curve (EAC) of ^{11}B. The curve MAR is the result of a constant rate of production by galactic cosmic rays (linear increase with time). (From Reeves et al., 1973)

VII. THE "RAINY" GALAXY

One extra complication may be awaiting us. Throughout these notes, the view has been implicitly taken that the galaxy is a closed system and that no appreciable amount of intergalactic matter has been added to the system This may be true or not. A few years ago the observation of fast-infalling clouds of possibly extragalactic matter (Oort, 1958) made the astronomical community alert to this possibility. Today, few people think that these clouds are really extragalactic. Nevertheless the "rainy" galactic model cannot be excluded (Quirk and Tinsley, 1972). It is reasonable to assume that the intergalactic gas is made of unaltered big bang nucleosynthesis, i.e. with the big bang content of D, ^3He, ^4He and ^7Li but none of the typical products of stellar evolution ($Z \geq 6$).

A continuous inflow of intergalactic matter in our galaxy would of course counterbalance the effect of astration on D, ^3He and ^7Li. Such a possibility has been considered by Tinsley and Audouze (1973).

VIII. CONCLUSIONS

Leaving aside these finer details of the theory, we may stress the important points in which we have a little more confidence: the atoms from D to B are produced in two main sources: the big bang and high energy nuclear reactions during the galactic life. It appears reasonable to assign D, ^{3}He, ^{4}He and a part of ^{7}Li to the big bang while the G.C.R. are responsible for ^{6}Li, ^{4}Be, ^{10}B, ^{11}B and a part of ^{7}Li. The exact origin of ^{7}Li is still unclear.

REFERENCES

Audouze, J., and Truran, J.W., Astrophys. J., in press,1973.

Audouze, J., and Fricke, K.J., preprint OAP-322, 1973.

Audouze, J., Lequeux, J., and Reeves H., to be published in Astron. Astrophys., 1973.

Baedecker, P.A., in Handbook of Elemental Abundances in Meteorites (Ed. Mason, B.),Gordon and Breach, New York, 1971.

Beer, R., Farmer, C.B., Norton, R.H., Martonchik, J.V., and Barnes, T.O., Science, 175, 1360, 1972

Black, D.C., Nature, 234, 148, 1971.

Boato, G., Geochim. Cosmochim. Acta, 6, 209, 1954.

Bradt, H.L., and Peters,B., Phys. Rev., 80, 943, 1950.

Burbidge, E.M., Burbidge, G.R., Fowler, W.A., and Hoyle, F., Rev. Mod. Phys. 29, 547, 1957.

Cameron, A.G.W., Colgate, S.A., and Grossmann, L., Nature 243, 204, 1973.

Cameron, A.G.W., and Fowler, W.A., Astrophys. J., 167,111, 1971.

Casse, M., and Goret, P., to appear in the Proceedings of the Denver Conference on Cosmic Rays, 1973.

Cesarsky, D.A., Pasachoff, J.M., and Moffet, A.T., Astrophys. J. Lett., 180, 1, 1973.

Cohen, J.G., Astrophys. J., 171, 71, 1972.

Cohen, J.G., and Gradsdalen, G.L., Astrophys. J.,151,L48,196

Colgate, S.A., Astrophys. J., in press, 1973.

Danziger, I.J., Ann. Rev. Astron. Astrophys., 8, 161, 1970.

Ezer, D., and Cameron, A.G.W., Canadian J. Phys.,43, 1497, 1965.

Fowler, W.A., in the Astrophysical Aspects of the Weak Interactions (Academia Nazionale de Lincei-Roma) p.116, 1971.

Fricke, K., submitted to Astrophys. J., 1973.

Geiss, J., and Reeves, H., Astron. Astrophys.,18, 126, 1972.

Ginzburg, V.L., and Syrovatskii, S.I., The origin of cosmic rays, Pergamon Press, New York, 1964.

Grevesse, N., Solar Phys., 5, 159, 1968.

Hall, D., to be published, 1972.

Hauge, O., Engvold, O., Astrophys. Lett., 2, 235, 1968

Herbig, G.H., and Woolf, R.J., Ann. Astrophys., 29, 593, 1966.

Hoyle, F., and Fowler, W.A., Nature, 241, 384, 1973.

Iben, I.Jr., Ann. Phys., 54, 164, 1969.

Jefferts, K.B., Penzias, A.A., and Wilson, R.W., Astrophys. J. Lett., 179, 457, 1973.

Kirshner, R.P., Oke, J.B., Penston, M.V. and Searle, L., preprint, 1973.

Lecce Conference on "Supernovae", May 1973.

Meneguzzi, M., Audouze, J., and Reeves, H., Astron. Astrophys., 15, 337, 1971

Meneguzzi, M., and Reeves, H., in preparation, 1973.

Merchant, A.E., Astrophys. J., 143, 336, 1966.

Mezger, P.G., preprint, 1973.

Michaud, G., and Vauclair, S., in preparation.

Oort, J.H., Inst. de Physique, Solvay Brussels, 11, 163, 1958.

Peebles, P.J.E., "Physical Cosmology" Princeton Series in Physics, 1971.

Quirk, W.J., and Tinsley, B.M., to be published, 1972.

Reeves, H., Astron. Astrophys. 215, 1972.

Reeves, H., Fowler, W.A., and Hoyle, F., Nature, 226, 727, 1970.

Reeves, H., Phys. Rev. D., 6, 3363, 1972.

Reeves, H., to appear in the Proceedings of the Cambridge Conference on Cosmochemistry (1972), 1973.

Reeves, H., Proceedings of the Lecce Conference on "Supernovae", to be published, 1973.

Reeves, H., Audouze, J., Fowler, W.A., and Schramm, D.N., Astrophys. J., 179, 9o9, 1973.

Salpeter, M., Astrophys. J., 121, 161, 1955.

Schmidt, M., Astrophys. J., 137, 758, 1963.

Schramm, D.N., OAP no. 275, 1972.

Schwarzschild, M., "Structure and Evolution of the Stars" Princeton University Press, 1958.

Searle, L., and Sargent, W.L.W., Comments on Astrophys. and Space Sci., 4, 59, 1972.

Shapiro, I.S., Astron. J., 76, 291, 1971.

Sill, C.W., and Willis, C.P., Geochim. Cosmochim. Acta, 26, 1209, 1962.

Smith, R.L., Sackmann, I.J., and Despain, K., in preparation, 1972.

Solomon, P.M., and Woolf, N.J., in press, 1973.

Spitzer, L., Drake, J.F., Jenkins, E.B., Morton, D.C., Rogerson, J.B., and York, D.G., to appear in Astrophys. J. Lett., 1973.

Tinsley, B.M., and Audouze, J., in preparation, 1973.

Trauger, J.T., Roesler, F.L., Carleton, N.P., and Traub,W., Meeting of the A.A.S., March 1973.

Truran, J.W., and Cameron, A.G.W., Astrophys. and Space Sci., 14, 179, 1971

Ulrich, R.K., and Scalo, J.M., Astrophys. J., 137, 1972.

Van den Bergh, S., Astrophys. J., 125, 445, 1957.

Wagoner, R.V., Astrophys. J. Suppl. No. 162, 18, 247, 1969.

Wagoner, R.V., Fowler, W.A., and Hoyle, F., Astrophys. J., 148, 3, 1967.

Wagoner, R.V., Astrophys. J., 179, 343, 1973.

Wallerstein, G., and Conti, P.S., Ann. Rev. Astron. Astrophys., 7, 99, 1969.

Weinreb, S., Nature, 195, 367, 1962.

Zappala, R.R., Astrophys. J., 172, 57, 1972.